JN424669

We want to design spatial affordances and conditions, conducive to our complex life.

Writer (저자)
(Design Group OZ: SeungSoo Shin, SangJin Lim, JaeWon Choi)
(㈜디자인그룹오즈: 신승수, 임상진, 최재원)
(SPACESPACE: Takanori Kagawa, Junko Kishigami)
(스페이스스페이스: 타카노리 카가와, 준코 키시가미)

Date of Printing (발행일) 30 March, 2013 (2013년 3월 30일)

Publisher (발행인) Kim, Ki-hyun (김기현)
Publishing Office (발행처) Spacetime (시공문화사)
5F Kukdong bldg., 200 Hyunjeo-dong, Seodaemun-gu, Seoul, Korea
(서울시 서대문구 현저동 200 극동빌딩 5층)
Tel. 02-3147-1212, 2323 Fax 02-3147-2626

Book Design & Editing (북디자인/책임편집) Yi, Soon-hak (이순학)
jade-comm@hanmail.net

ISBN 89-5592-259-0
25,000 KRW

Thus we approach each project to make a complex organization rather than complicated one, by superposing multiple design value on a place and space: especially on a dead space, a common space, an extra space, and a remnant space.

共存
方式

THE WAYS OF BEING TOGETHER

ACTIVATING BOUNDARY

디자인 그룹 오즈
Design Group OZ

INTRODUCTION | 여는 글

이 책은 지난 11월 16일부터 12월 9일까지 토탈 미술관에서 열렸던 한일현대건축교류전의 결과물 가운데 하나다. 우리는 이 전시를 준비하는 동안 일본측 카운터파트너인 SPACESPACE와 'the ways of being together' 라는 소주제를 가지고 여러 가지 이야기를 나눌 기회를 갖게 되었다. 이러한 토론을 통해서 이제까지 우리가 추구해왔던 디자인의 방향을 더욱 더 명확하게 규정할 수 있게 되었고, 우리와 일본측 전시자인 SPACESPACE 사이에 존재하는 차이점과 공통점도 보다 잘 이해할 수 있게 되었다.

우리가 이 책에서 말하고자 하는 것은 우리가 생각하는 공존의 방식이며, 이것을 어떻게 디자인할 것인가에 관한 것이다. 부제인 'un-private house' 는 바로 이 문제에 대한 우리들의 생각이 담겨진 핵심적인 키워드라고 하겠다.

우리는 건축이 지속적인 여지(room)의 집합이라고 생각한다. 여지란 어떠한 사건(event)이 일어날 수 있는 가능성이며, 사건과 활동을 통해서 장소를 점유하는 것이다. 그런데 건축은 어떠한 사건보다도 오랫동안 장소를 점유하는 가장 지속적인 사건이다. 그래서 건축은 사적인 공간일때조차도 공공성을 갖는다.

또한, 우리는 공공성이 개인의 삶을 떠나서 존재하는 것이라고 생각하지 않는다. 공공성이란 개별적 삶들 사이에 존재하는 공유의 공간, 그리고 공존의 공간이라고 믿기 때문이다. '세상 풍경 중에서 가장 아름다운 풍경은 모든 것들이 다 제자리로 돌아가는 풍경' 이라는 노랫말처럼, 일상의 삶과 삶의 자원들이 제자리를 찾는 모습이야말로 지속가능한 사건의 모습이며 공존의 풍경일 것이다.

따라서 우리는 '공존의 방식' 을 '사적이지 않은' (un-private) 무엇에서부터 찾고자 하였다. 이것은 사적인 동시에 공적인 영역이며 개별적인 동시에 집합적인 영역인 사이의 공간, 즉 '경계 공간' 이며 '전이의 공간' 이다. 우리의 디자인 작업은 바로 이 경계의 공간을 조직함으로써 사건과 활동의 여지를 만들어가는 것이다. 따라서 우리의 작업은 관계지향적이고 작동적이며(operational), 깊이 있는 경계를 지향한다. 마치 여러 가지 물건들을 보자기로 싸듯이, 우리는 여러 가지 사건들과 활동들을 깊이 있는 외피로 묶어서 한 다발의 이야기 꾸러미를 만들어낸다.

한편, SPACESPACE의 디자인 접근방식은 작동적이라기보다는 환경적(environmental)이라고 할 수 있다. 이들의 공간은 깊은 경계를 만들기보다는 딱딱한 경계를 지워나가는 방식으로 공존을 이야기한다. 이 공간 속에 자연 혹은 환경은 하나의 대상이 되어 건축 공간과 관계를 맺으며, 이들 사이의 경계는 모호해지고 얇아진다. 이런 측면에서 Design Group OZ가 인간과 인간 사이의 상호작용을 주제로 작동적인 공존방식에 중점을 두고 있다면 SPACESPACE는 인간과 사물 사이의 상호작용을 주제로 환경적인 공존방식에 초점을 두고 있다고 말할 수 있다.

이와 같은 차이점이 있지만, Design Group OZ와 SPACESPACE 모두는 대규모 아파트 단지로 대표되는 오늘날의 획일적인 도시 주거 형식에 맞서서 장소성과 일상성을 회복하고 다양한 가치가 공존하는 집합형태(collective form)를 실험하고 있다는 점에서 공통점을 가지고 있다.

우리는 공존의 디자인, 공공의 디자인이란 무엇보다도 개인을 사회와 소통시키는 공간조직이라고 생각한다. 이를 위해서 먼저, 다양한 개별적인 관심과 활동을 촉발하고 선택의 폭을 넓히는 디자인, 수동적인 경험 대신에 능동적인 참여를 유도하는 비우는 디자인, 그리고 개별적 공간을 연계하여 도시로 열어가는 디자인이 요구된다. 이러한 공간조직은 경험의 강도만을 강조하는 환상 도시로부터 경험의 빈도와 지속성에 무게중심을 두는 공유 도시를 향한 우리들의 비전과 이야기를 담고 있다.

말풍선 속에 담긴 우리들의 작은 이야기와 공간들이 모여서, 커다란 도시의 비전과 시나리오가 만들어지듯이, 우리가 꿈꾸는 공존의 방식들도 우리 주변의 개별적인 사건들의 새로운 집합방식으로부터 자라나고 가꾸어진다는 것이 이 책이 궁극적으로 전하고 싶은 이야기다. 이것은 un-private이라는 단어처럼 익숙한 동시에 낯선 것으로, 일상 속에서 재발견된 새로움이라고 할 수 있다. 그리고 이것은 개별적인 정보(in-formation)를 열어서 혁신적인 가치 질서를 만드는 과정, 즉 형태화(formation)이며 우리가 지향하는 디자인의 방식이다. 깊은 경계 속에서 안과 밖이 화해하는…

신승수, 임상진, 최재원 / ㈜디자인그룹오즈

INTRODUCTION | 여는 글

This book is an output of Korea-Japan Architecture Exchange Exhibition, held from November 16 to December 9, 2012 in The Total Museum of Contemporary Art. While preparing the exhibition, we had an opportunity to discuss various subjects on our topic of 'the ways of being together' with our Japanese counter partner, SPACESPACE. With this discussion, we came to define the design direction we have pursued more clearly and understand the difference and common ground between the two more distinctly.

we came to define more clearly the design directions we have pursued and understand more distinctly the differences and similarities between the two.
There are the ways of being together that we think and the ways of how to design these, which we want to talk about in this book. In this sense, the subtitle 'un-private house' would be a crucial keyword for expressing our ideas on this subject.

Architecture, we believe, is a sustainable assemble of room. The room is a possibility of certain events to take place, and it actually takes 'place' through events and activities. In this way, architecture is the most sustainable event that takes 'places' quite longer than any other event. Thus, architecture holds publicness even when it belongs to private property.
Architecture, we believe, is a mass of sustainable rooms.

Furthermore, we do not think publicness does exist beyond individual life, as we believe that it is nothing but a space of sharing and coexisting, between individual lives. Like a lyric says 'The most beautiful landscape in the world is the landscape which lead us to the right place', it is the landscape of being together that makes everyday life and its resources at the right place, and that makes events truly sustainable.

Thus, we have approached 'the ways of being together' from something 'un-private', which is a space of in-between, a territory of being private and public, individual and collective at the same time, and the space of 'border' and the space of 'transition' in other words. Organizing this space boundary, our design work is to make room for events and activities. Therefore, our job is relations-oriented, operational, and aiming for a deep boundary. Just like wrapping many things in a

cloth, we intend to make a bunch of stories by binding various events and activities together with a deep skin.
While, the design approached by SPACESPACE is to be environmental rather than operational. Their space talks about the ways of being together by removing hard boundary rather than making deep one. Within this space, the nature or the environment turns into a spatial object and enters into such a close relation with architecture that the boundary between them becomes much more ambiguous and thinner. In this sense, we might say that SPACESPACE is focusing on the environmental coexistence regarding the interactions between people and things, whereas Design Group OZ is placing emphasis on the operational coexistence regarding the interactions between people.

Despite these differences, Design Group OZ and SPACESPACE share common ground. Both of them have been searching for the sense of place and everydayness in their design, and have been experimenting collective forms where various values are being together, against today's uniform types of urban housing, typified by massive apartment blocks.

The design of coexistence and publicness should be a spatial organization that enables individuals to communicate with society. For this, a set of consistent design approaches are needed, such as a design, offering a wide choice of spatial engagement and activating multi-layered individual interests and activities, a void-generating design, inducing active participations rather than passive experiences, and a design, connecting individual spaces and openning them to the city.

What we intend to convey in this book is 'the ways of being together' that we have dreamed of can be grown and cultivated from a new set of individual events around us, as our small story and space in the word balloons gather together to make a great vision and a scenario of a city. This is the rediscovered newness in a daily life, as if the word of 'un-private' is familiar and unfamiliar at the same time. And it is the process, so called 'formation', which opens an individual information(in-formation) to make an innovative value-system. This is what we aim to design, bringing the inside and the outside back together in a deep boundary...

SeungSoo Shin, SangJin Lim, JaeWon Choi / Design Group OZ

RECOMMENDATION | 추천의 글

공생과 배려에 의해 생성되는 공공성

경희대학교 김일현

공공성과 집합성

요즘처럼 도시를 논하면서 공공성에 대해 논의가 활발하게 이루어진 적도 별로 없다. 이전까지 위에서 아래로 구색을 갖추는 수식어로 그 단어는 사용되었고 다수의 시민은 무관심으로 이에 대응해왔다. 공공이나 소통과 같은 단어가 빈번하게 사용되지만, 밖을 향한 파장보다는 용어 내부에서 귀결되는 공허한 공명이 더욱 두드러진다. 친숙한 용어를 반복하다 보면 어색해지듯이, 그리고 지금처럼 양적 팽창과 성장위주의 개발과 발전이 객관적인 위기의 상황을 직면했을 때, 비로소 진지하게 본연의 의미를 성찰할 수 있는 여건이 주어진다. 반복되는 사전적 의미가 공허한 이유는 아마도 암묵적으로 우리의 내면과 사회의 관습으로 잠재되어온 유교문화와 급작스럽게 수용한 근대주의의 영향 때문일 수도 있다. 하지만, 가장 큰 이유는 진지하게 이에 대해 논의할 시간과 기회가 없었기 때문일 것이다. 이러한 상황을 서양에서 통용되는 용어와 사례를 인용하는 것으로 무마하는 경우도 빈번했다. 하지만 이론보다 현실이 보다 역동적이기에, 이제는 구체적인 현상과 실재를 존중하며 이에 맞는 용어를 파악하는 것이 필요할지도 모른다. 지금까지 과거를 돌아볼 여유조차 없었지만, 양적 공급에 근거했던 비대한 건설시장의 위기와 맞물린 현 시점에서 광장, 공원 그리고 거주의 공공성에 대해 직접적인 질문을 할 시점에 준비도 되지 않은 상태에서 도달했는지도 모른다. 의미의 해석은 다양하겠지만, 명백한 점은 위의 용어들을 절박하게 반복할 만큼의 단절과 균열이 사회내부와 공간구성에 만연하기 때문일 것이다. 혹자는 시민사회의 등장과 함께 탈 개인화 된 공공성의 재현을 논하며, 현대사회에서 개체의 집합이 구성하는 영역성과 이를 중재하는 상위 행정권력의 역할을 주장하기도 한다. 하지만 이러한 전제로 실행되는 정책은 개인과 권력의 복잡한 이해관계라는 변수가 원칙을 압도하는 경우가 빈번하다.
항상 원점에서 공공성을 논의하는 듯 하지만, 적어도 이제는 회피할 수 없는 현실자체를 전제로 논의할 수 밖에 없으며, 이 구도 속에서 계획과 이론은 반영과 투사라는 실재와의 긴장관계 사이에 위치하게 된다.

도시와 거주

세계인구의 절반 이상이 도시에 거주하고 있으며, 우리의 경우 수도권과 광역시 그리고 주요도시에 거주하는 인구의 밀도는 세계의 평균을 훨씬 웃돈다. 환경이나 에너지 효율의 측면에서 소모적이며 돌발적인 문제들이 존재하지만, 비례할 정도의 매력과 중독성이 있어서 쉽게 떠나지 못하는 곳이기도 하다. 빈틈이 없을 정도로의 밀집된 상태에서 인접한 거주자들과의 접촉은 정주성의 기간에 비례하여 이루어진다. 즉각성의 매력에 집결해서 도시에 거주하는 익명의 대중과 개별적 운명들은 서로 우연히 교차하고 동시에 분주하게 분할된 시간을 접속으로 채운다. 이처럼 풍요에 대한 미래의 희망과 여유에 대한 현재의 갈망이 교차하며 도시는 이에 대응하는 수 많은 장치들로 채워지고 있다. 서울을 비롯한 우리의 도시들은 세계의 어느 곳보다도 이러한 역동성을 보이며 명분과 이해 사이에 오가는 격렬한 관계를 표출한다. 이처럼 도시는 단순한 건조환경이 아니라 거주하는 사람들의 분신과도 같다. 이러한 영역 속에 위치하는 집은 고정된 위치를 점유하지만 세대수뿐만 아니라 거주자의 연령, 즉 인생주기에 의해 계속 변화한다. 그리고 현 상황과 시간에 대한 인식은 가장 기본적인 질문, 과연 이렇게 사는 것이 최선인가라는 삶의 질에 대한 문제의식에서 시작된다.

팔라디오의 빌라에서 브라만테의 도심형 주상복합 건물, 르 꼬르뷔제의 집합주택에서 아이젠만의 주택시리즈에 이르기까지 삶의 방식을 담는 주택은 현실과 이상이 공존하는 전형적인 매체의 역할을 수행했다. 유럽의 경우, 제2차 세계대전 이전에 생산과 거주의 통합이라는 주제는 전쟁 이후 파괴된 도시의 재건 그리고 양적 공급을 목적으로 추진되어 왔다.
공공성에 대한 논의는 유사한 단어를 사용하면서도 계속해서 그 내용은 급변해왔다. 생활공간과 공존하는 공간에 대한 관심은 특히 1960년대에 두드러졌다. 그리고 최근 네덜란드와 스페인의 공공주택과 단지계획은 점차적으로 무게의 중심을 유형에서 집합으로 전이해왔다. 성장위주의 개발이 한계에 부딪히고, 슬럼화된 고층집합주거라는 현실을 직면하며, 더 이상 도시가 단기적 투기와 무마의 대상이 아닌 삶의 현장이라는 교훈을 값비싼 시행착오를 통해 공감하기에 이르렀다. 또한 막상 모여서 무엇을 개선하려고 하는가, 집합성의 의미는 무엇인가를 논하면 명쾌한 답은 떠오르지 않는다. 아마도 이는 집합 이전에 중요한, 개인에 대한 논의가 심층적으로 진행되지 않았기 때문일 것이다. 공공성은 배려의 산물이다. 억지로 공유할 수 없을뿐더러, 공간을 할애한다고 소통이 진행되는 것도 아니다. 스미슨 부부의 로빈 후드 가든에서 여러 건축가들의 한자피어르텔(Hansaviertel)의 계획, 혹은 소비에트 구성주의의 '소셜 컨덴서'와 같이 건축가의 일방적인 제안이나 지자체의 강압에 의해 공공성이 작동하지는 않는다. 다행스럽게도 이러한 실험의 사례들과 사회통합의 도구들이 거대이념과 담론이 배제된 전제에서 재고찰되는 시점에 이르렀다.

공생의 수거

우리에게 유형으로써의 공동주택은 친숙하지 않다. 오히려 마을 혹은 공동체가 익숙했지만, 지난 50년간 압축적으로 경험한 변화속에서 많은 가치들이 상실된 것은 객관적인 사실이다. 그리고 집에 대한 관념도 삶의 공간이기보다 부를 축적하는 수단이었고, 성인인구의 대부분이 실물경제의 전문가인 만큼, 재태크와 교육에 있어서 집이 어떤 의미를 지니는 지에 대해서는 다 강의를 부탁하면 한 시간정도는 이야기를 할 수 있을 정도일 것이다. 하지만 막상 본인이 어떤 집에서 살고싶으냐에 대한 질문은 자신과 먼 상상의 영역으로 다가온다. 도시는 변화하는 속성을 근본적으로 지니며, 그 내부에 위치한 건물들도 시간이 지남에 따라 성격이 바뀐다. 비근한 예로, 과거의 단기간 개발의 방식이었던 도심형 한옥이 이제 문화유산으로 도시에 개성을 부여하는 상황에서 대도시에 편재한 주상복합시설이나 다세대주택에 대해서도 기존의 건축과 건물의 분류 방식이나 백지화의 대상이 아닌 내재적 가치를 인정하는 시선으로 바라볼 필요가 있다. 3p house 봉천동이 위치한 주변은 바로 그러하다. 모더니즘이 형식주의로 전락하고, 정치권력과 자본사이에서 갈등하던 시기에 건축가들은 작품으로써의 건축이 아닌 일상의 건물에 눈을 돌렸다. 우리나라의 다세대 주택은 이런 의미에서 매우 도시를 상징하는 대표적인 유형임이 분명하다. 최근에 와서 젊은 건축가들을 중심으로 이러한 건물로 구성된 지역에 대한 관심 그리고 새로운 접근의 제시가 드러나고 있다. 어쩌면 도시경관의 대부분을 차지하는 건물을 집장사나 허가방의 산물이라고 폄하하는 자세에서 이를 적극적으로 수용하며 가치인식과 작업의 범위에 대한 태도가 변했다고 볼 수 있다. 이 경우 거주의 범형에 대한 연구와 고찰이 두드러진다. 단순히 하나의 유형을 제시하는 것이 아니라 조합을 통해서 또 다른 집합이 구성되는 방식을 모색하게 된다. 건축가는 대지를 선택하지 못할 뿐만 아니라 주변의 경관도 선택하지 못한다. 그러므로 외부와 접하는 방식에 대한 전략을 수립할 수 밖에 없는데, 봉천동주택의 경우 창을 통한 다양한 풍경과의 관계 맺기를 모색하였고, 이는 건축집단이 밝히듯 추상적인 방식으로 전통가옥의 개구부를 해석한다. 익명적인 도시에서 공존할 수 밖에 없는 조건이 거주환경으로 주어지지만 은밀한 공간을 소유하려는 필요성도 다양한 방식으로 추종한다. 필지에 의해 단절된 통로의 공유라는 건축가의 도시에 대한 입장을 수용한 건축주들의 배려가 있었고, 순차적 배려를 통해 외부공간의 주민과의 공유, 개별 중정의 제공, 그리고 마지막으로 입주자들만 공유하는 옥상정원이 놓여졌다. 외벽의 일체화를 통해 여러 유니트를 하나로 통합시키려는 의지가 보인다. Extruded Community 문정동 공공주택의 경우 공동의 소유가 공동의 책임을 시사하고, 진정한 의미에서의 공공은 참여에 의해 완성됨을 보여준다. 벽을 높이면 역설적으로 더욱 안전해지기 보다는 고립되는 주거공간이 반증하듯, 여럿이 공유할수록 외부의 위협에 공동으로 대응할 수 있다. 이는 기숙사나 원룸과 같이 통합된 듯하면서도 결국 개체로 분할되거나, 고시방이나 쪽방과 같이 파편화되고 균열되며 단절된 내부의 단위를 양산하는 것과는 다른 선택이다. 이렇게 자발적이고 자율적인 공간의 운영, 일시적으로 나마 공동체의 동질성을 갖는 집합적인 주인이 있어야 공간은 유지된다.

OZ는 논쟁적인 집합성(collectivity), 공공성(publicness), 그리고 다중성(multiplicity)과 같은 용어에 함몰되기보다는 보다 유쾌하고 일상적인 방식으로 '함께' 혹은 '더불어 살기'로 명명하면서, 공공성이나 집합성 등의 단어가 기표로써 지니는 부담감과 거리를 취한다. 즉 강제로 한 장소를 함께 점유하도록 하기 보다는 가진 것을 공동으로 소유하고, 그 소유집단의 정도를 조절하는 방식을 취한다. 이러한 입장을 표명하기 위해서 보다 적절한 표현의 방식이 요구되는데, 대중과의 거리를 없애기 위한 일환으로 비야크 잉겔스와 비슷하게 경쾌하고 친숙한 만화의 형식을 활용한다. 뿐만 아니라 현재 입장에서 잉겔스는 롤모델이나 스파링파트너와 같은 역할을 수행한다. 이러한 양상은 특히 '거리주택'(Street Housing)에서 두드러진다. 일종의 수평 거대구조의 형식으로 최대한 수직구조의 단점과 획일화되는 집합주거의 조합을 지양하고자 하였다.

동시에 이 계획안은 B.I.G의 VM Mountain in Ørestad와 Architekten Cie의 암스테르담 집합주거인 '고래'에 대한 일종의 오마쥬로 보인다. 경사를 주된 모티브로 삼은 이 집합주택은 산책로의 다른 해석을 하며 가로를 데크가 아닌 방식으로 내부로 끌어들인다. 가로의 삽입을 통해 각 층을 부유하는 대지로 인식시키려는 전략을 적용하는 동시에, 경사에 대응하는 개별 유니트의 개체성이 확보되는 방식의 조합을 적용하였다. 최근, 이러한 산책로를 보다 내부로 삽입하는 동시에 외부로 확장하려는 일련의 시도들이 여러 건축가들을 통해 진행되었다. 바닥에 대한 고정관념에서 벗어난 듀플렉스 유형의 탐구는 내부경관 혹은 랜드스케이프의 다양한 시도로 확장된다. House crossover에서 약간의 변화로 다양해질 수 있는 사람의 행태와 공간의 활용방식, 그 의의는 백지수표가 아닌 제약된 조건에서 더욱 빛을 발한다. 꽈배기와 같은 유니트의 반복을 통해 최대한의 활용면적을 확보하는 동시에 함께 공유하는 지하공간과 옥상공간를 할애하면서 아파트와는 차별되는 집합성의 가능성을 제시한다. 아쉬운 부분은 통합된 단지에서 또 다른 전체를 구성하는 얼개가 잘 두드러지지 않는다는 점이다. 역사적 선례로는 르 꼬르뷔제가 제시했던 프랑 오부스(Plan Obus) 와 도미노(Dom-ino)의 다양한 조합방식 에 대한 재해석이 엿보이고, 보다 구체적인 유사관계는 판덴브로크와 바케마(van den Broek & Bakema), 그리고 니마이어(Niemeyer)의 베를린 주택을 연상시킨다. House Cornerstone에서는 건축가들이 대면할 수 밖에 없는 사선제한이라는 건축한계선을 기회로 삼아서 각층마다 테라스를 부여하는 방식을 취했고, 반복과 변화를 통해 평이할 수도 있는 띠창을 하나의 leitmotif로 만들다. 동시에 통일된 벽돌이라는 재료를 통해 critical mass를 구현하고 단일한 형체로 기념성과 인지성을 높이고자 한듯하다. 이와 다른 위치에 놓은 프로젝트에서는 수려한 자연환경을 맥락에 이중적인 자세로 대응한다.
Street pool villa에서 주된 주제는 극장성인데, 대지와 건물의 통합하여 인공적인 기단을 조성하는 동시에 분할된 덩어리 사이로 변화하는 풍경을 끌어들인다. 그리고 사선으로 구성된 면으로 긴장감을 연출하고 수려한 자연과의 중재가 이루어진다. 한편으로 자연과 인공의 대비를 극대화했던 마이클 하이저(Michael Heizer)의 구조물을 그리고 다른 한편으로 이사무 노구치(Isamu Noguchi)가 시도했던 놀이터의 지반을 연상시킨다. 균질한 기하학을 전제로 하여 퍼즐을 파내는 듯한 방식으로 구성된 개별 공간은 실질적으로 모든 활동을 담는 이벤트 프라자에 의해 통합된다. 마지막으로 반대급부에는 도심지 내에서 동양적 공간을 형이상학적으로 모색한 Public garden house이 있다.

OZ의 일련의 계획안들은 유쾌한 방식으로 그 긴장관계 속에서 다양한 방식으로 공간의 설정과 점유의 방식을 실험한다. 고무적인 점은, 집요하게 함께 살아가는 집합적 거주에 대한 OZ의 연구이며, 일종의 공생을 전제로 하는 변주로 보인다. 이처럼 오즈가 공공성보다 공존을 내세운 것은 이러한 한국적인 상황을 반영한다. 그들은 깊은 경계, 즉 경계의 두께에 주목한다. 쌈지공원, 텃밭 혹은 통로와 같이 다양한 방식으로 그 두께는 다양한 모양을 취한다. 변화무쌍한 상황에서 건축은 그 자체로 완결체가 아니며 사람들이 점유하는 방식에 따라 다른 색채를 띠는 이벤트의 속성을 지닌다.

RECOMMENDATION | 추천의 글

"Publicity" generated by coexistence and the consideration.

Kyunghee University Ilhyun Kim

Publicity and Collectivity

There has never been a time in which such an active discussion about publicity occurred like nowadays. The term, "publicity" is overwhelmingly being used rather meaninglessly to decorate in rhetoric way by politicians and communication engineers to fill emptiness of their phrases and discourses. This causes most people to get numbed by the usage and to distance from essential and critical stance of the term. As words such as communication and publicity repeatedly being used, resonance into the internal void increases while impact toward the outer reality notably deteriorates. Current crisis caused by development and progress solely based on the quantitative expansion, also lead to a moment of reflection on the etymological and essential meaning of the term such as publicity and collectivity. Repetitive use of the terms might sound devoid of meaning has multiple cause, such as inherent Confucian habit as well as radical industrialization and its after math. However main and obvious reason probably is the lack of time and chance to discuss. As always, the reality overwhelms any theory; hence specific analysis is required not to castrate its vibrancy within empty words, specially adopting cases and neologism from Western thoughts and theories. If lack of time or self-esteem was characterized in the field of the debate, while the construction and real estate market have enjoyed until recently flourishing speculation. Extremely functional and short-term profit exceeded the rest of considerations for the public amenities and other welfares. Now the time is ripe to confront the inevitable changes that are occurring in metropolis and to raise questions on the public status of the places such as plaza, park and street and on the meaning of publicity of collective living in the city. So the terms might be interpreted in various ways, but current repetitive usage of those words shows how discontinuity and cracks within the society and its spatial configuration are present in actual time. Some might recall de-individualization with the advent of civic society and others the role of administrative power to converge the diverse interests of a collective formed by those individuals. However we have witnessed how complex interest and even conspiracy of mutual benefit turn common sense into common place. Seems like the discussion on the publicity seems to repeat 'ritornello', but this time, the consideration of reality cannot be ignored and in this circumstances the design and theory relocate themselves in tension with actuality with their reflective and projective characters.

City and Inhabitancy

Approximately half of world's population live in urban environment, and for example, huge ratio of Korean population resides in capital cities or satellite city, which is overwhelming the above average ratio of world's population. According to the energy and environmental issue, a city results inefficient and also contains unexpected problems, but its inhabitants choose to remain because of convenience and attractiveness or even dependency. Anonymous public residing in the city and the

fate of the individual is filled with connection of cross into each other, and at the same time fascinated immediacy, busy splitting the real-time with clicks into hyperlinks. Likewise, a city is filled with many devices corresponding to the crisscross of the hope of the future and abundance. Megalopolis like Seoul filled with these dynamics compared with any other places displays exasperation of connectivity between justification and comprehension. A city is not characterized not only by its physical entity but also with its inhabitants. Its placeness changes constantly as the residents age in their unique life cycle. Paraphrasing an old Italian anecdote says, "you are what you eat", as similarly Orhan Pamuk once mentioned about inseparableness between city and its dweller, we might assert "a city is alter-ego of its collective residents". Consciousness of the current situation and the time begins from the basic question which is awareness of issues about the quality of life.
From Palladian villa to Bramantesque urban complex buildings, from Le Corbusier's housing to Eisenman's house series, a house as an ideal-type has always been a medium in which the real and the ideal coexisted in conflict. In case of Europe, before and after the World War II, the argument of integration between production and residency was central issue until the reconstruction phase of the Fifties. Discussion about the publicity continued in similar terms but in totally different way, especially around the Sixties with the advent of movements so called "international utopia". Recently, public house and site plan in the Netherlands and Spain have been gradually transited focusing from type to diagrammatic set. As the growth-oriented development reached its peak and the degradation of high-rise collective housing results undeniable, a hard lesson had been learned that the city worth to be dealt not anymore as a jungle of speculation but with a long term consideration of common wealth. Still, there is no clear answer for such a simple question "what are we trying to improve" or "what the collective meaning of space is?" This is because it did not proceed for deep discussion about person importantly before dealing with set. Publicity is a product of the mutual care, hence forced sharing of certain physical space does not automatically leads to publicity since it requires adequate attitude beforehand. Publicity involves diverse participation and cannot be set by authority of local government or intention of a single architect; to mention a few of many examples, buildings by different architects' at Hansaviertel in Berlin, Smithson couple's Robin Hood Garden or 'Social Condenser' of Soviet constructivism. Fortunately, the case of experiments and the tool of the social integration are now thoroughly reconsidered with exclusion of grand discourse and ideology.

Houses of Coexistence

Co-housing is not a familiar culture in the recent Korean contemporary life hood. In the last three decades, the relationship based on the genealogy and 'big family' had been reduced to the minimum season reunions. Sense of community had been altered as well, as dwellers feel more comfortable with online buddies with similar hobbies and interest rather than next door neighbors. Notion of home or shelter had been substituted with that of speculation and accumulation of wealth, turning most of the adults into experts of real-estate investment. The city has fundamentally altering character of the properties as buildings located within undergo parallel changes constantly. For example, urban Hanok once was an economical investment strategy of the short-term real estate speculation in the past but now a cradle of value of cultural heritage. Likewise, there is a need to look a residential complex or multi-family housing at the viewpoint of the intrinsic values rather than thorough pre-established mind set or as an object of tabula rasa.

3p House located at Bongcheon area deals this issue. As modernism in architecture has been degenerated into formalism devoid of semantic contents, architects turned to the building not as an architectural object but as a whole of co-respective parts to the daily life. Anonymous multi-family housing not only quantitatively but qualitatively is representative type for city, and in recent years it became object of interest for unbiased young architects. These buildings that occupy most of the urban landscape are now seen not as leftover of previous real estate development, but as embodiment of intrinsic value that requires alternative approach and attitude. 3p House is in this sense a case study which opens a discussion on the collective type of housing. It seeks another set of configurative combination through mixture rather than simply suggesting function per type. As architects usually cannot choose the site for the destination of a building but take it for granted, this building reflects under certain limit the strategy how to contact with the outer ambience. Bongcheondong house found the way of relating with a variety of urbanscape through a set of windows, similar to the fenestration of the traditional Korean houses in an abstract manner, as mentioned as well in the brief. Concern to enhance intimacy of the residency is considered within an inevitable condition where the building is destined to be located within an anonymous urban context. Site condition of the lot that was meant to divide front and rear areas were connected thanks to the acceptance by the client the proposition of architects. At the same time through gradual consideration, outer space has been shared with the neighbors, inner court given to each resident hood and roof garden shared by whole resident community. To enhance the sense of the integrity, multiple units were unified with a single outer wall and its cladding.

The work of Extruded Community of Munjeongdong suggests that common ownership is collective responsibility. It also shows that the true meaning of public is completed in participation. Gathered units without integration paradoxically become threatened and isolated as fence becomes thicker and higher. Sharing space grants common identity to the residents and to collectively correspond to the outer factors. It is a position different from dormitory or studio type officetel, where units are juxtaposed but not unified, nor the Gosiwon or dosshouse where internal fragmentation and disconnection are more intensely present. The space becomes managed and maintained with the presence voluntary and operations of autonomous space by a collective owners sharing identity of community even though it is temporary.

DESIGN GROUP OZ makes much importance of 'together' or 'live together' by pleasant and inviting manner rather than being depressed on terms of controversial collectivity, publicness and multiplicity, also they intentionally keep distance from burden of signifiers of those terms. In other words, they do not pick occupied a place by force but take the way of cross-ownership and management of them. It is required to more appropriate expression due to support the fact. They utilize playful and friendly cartoon format like Bjarke Ingels to close the distance with people. Moreover, they look at Ingles and apply for as a role-model or sparing partner. These aspect stands out especially in 'Street Housing'. It keeps a sort of

horizontal large-scale structure but avoids a weakness of maximized vertical structure and a combination of uniformity housing. In the same time, this proposal seems to confront similar context of VM Mountain in Ørestad by B.I.G and The Whale in Amsterdam by Architekten Cie. Street Housing which has as the main motif of slope comprehends a walk in another direction and lead it to interior in different way not as an additional deck but as an alternative and suspended ground. The strategy is to enhance an awareness of each floor wealthier through insertion of a horizontal and to apply combination method which is secured the individuality of every unit corresponds to the slope. Recently a group of Dutch architects proposed the projects that the path is inserted in the inner space and extended to exterior. Abandonment of pre-established notion of the floor and consideration of potential of duplex type allow diverse experiments of combining inner scenery with urbanscape in theatrical way. House crossover intends to make lots of varieties as the behavior of people and take advantage of the space into slight changes. It has a value in the constrained conditions rather than non-planed. Space is made the most of utilized area through the repetition of the units, such as the pretzel. It is presented a possibility of collectivity which is differentiated with apartment to share the underground space and rooftop simultaneously.
Unfortunately, it does not remarkable well just another full integrated configuration structure. It is shown re-interpretation of various combinations that Le Corbusier suggested in Plan Obus and Dom-ino. More specific similarity relationships are Berlin housing of van den Broek & Bakema and Niemeyer. Architecture is faced to limited construction line that it is considered necessarily for planning in House Cornerstone. As a good opportunity, house Cornerstone is provided terrace on each floor and made the leitmotif of one of horizontal window through repetition and change. Furthermore, it made critical mass through unified brick materials, and it seems to improve a memorial and recognition as a form. The project in a different location copes with beautiful natural environment in a double meaning. The main theme from the Street pool villa is theatricality. It drags the changing landscape between divided mass at the same time, creates an artificial podium in the integration of land and buildings. The diagonal façade directs a tension and it is taken arbitration with nature comfortably. On one hand, it reminds of Michael Heizer's structure which maximizes the contrast of natural, and on the other, Isamu Noguchi's playground. An individual space consists of a manner of pulled out a puzzle based on homogeneous geometry. It is incorporated by event plaza which holds substantially all of the activities. Finally Public garden house is an example to seek an oriental space metaphysically in urban areas.

Design Group OZ investigates to a variety of ways in pleasant how the space is occupied and set up in the tensions. They research a collective housing in living together intensely based on symbiotic relationship. To advocate not publicity but coexisting for Design Group OZ is a result to reflect the Korean situation. They are concerned about relationship of deep boundary and pay attention to the thickness of the boundary. Ssamzie park, garden or pathway and in a variety of ways, the thickness of the limits takes a variety of forms. Architecture by itself is not a finalized object, it is located within the ever-changing situation where its character also constantly changes as narrative events by the way that people occupy the space.

공존의 방 식

작동하는 경계

"공공 공간은 열린 공간이거나 접근가능한 공간으로 주어진 것에 불과한 것은 아니다. 마찬가지로, 사적인 공간 역시 닫힌 사유공간이 아니며, 두 공간 모두 개별적인 경험이 타자들의 경험과 함께 점유하고 있는 장소와 장소 사이에서 만들어진다."

방벽 도시

한국의 아파트 단지나 서구의 게이티드 커뮤니티(gated community)로 대표되는 현대의 파편화된 도시 공간은 개인화와 도시화, 그리고 유동화가 촉발한 다양하고 이질적인 헤테로토피아적 공간 속에서 다시금 근대적 의미의 공동체적 이상향을 꿈꾸고 있다. 그러나 이들 공동체는 어디까지나 균질한 가치로 채워진 내부지향적 공간들로서 이들 공간에서 살아가는 사람들은 같은 사람들을 만나고, 같은 행위와 같은 경험을 하도록 요구 받으며 같은 계급의식을 느끼고 다른 사람들의 현존을 지워내고 있다.

그들이 여전히 모이고, 만나고, 소통하고 있을지는 모르지만, 제한된 커뮤니티 속에 파편화되거나 정해진 동선 속에 갇혀 유리되는 삶을 산다. 더욱이 이들 균질한 가치의 방어벽이 쳐진 공간(Barricaded space)은 현대 사회에서 관찰되는 수 많은 미분화된 삶, 수 많은 층위에 걸쳐 존재하는 삶의 장소들을 끊어냄으로써 다양성의 충돌로부터 발생되는 창조적 가능성을 원천적으로 봉쇄하고 있다.

아파트의 거주면적별로 사회적 계급의식이 만들어지거나, 아파트의 건설사의 브랜드별로 공간의 가치가 결정되거나, 방음벽과 옹벽으로 둘러 쌓인 성곽 같은 공간이 선호되는 현상 등은 바로 우리 사회에 존재하는 보이지 않는 벽들의 단편들이라고 할 수 있다.

실제로, 한국의 대표적인 주거 공간인 아파트 단지는 말 그대로 사사성의 극단이다. 극단적인 사적인 공간을 만들기를 거듭한 결과 이들 공간에는 사적 공간과 공적 공간 사이에 존재하는 반-사적 공간과 반-공적 공간이 극단적으로 축소되거나 소거된다. 반면에 억제되지 않은 이질적인 욕망과 행위들이 주거공간 담장 너머에 만연해 있다. 하지만 이것들 역시 수많은 종류의 방과 상징물, 그리고 장치들 안에 포획되고 사유화되어 있기에 도시의 공공적 삶을 내부화시키고 우리를 타자로부터 분리시킨다.

사이-장소를 만들기 위해 경계를 작동시키다

우리가 제안하는 프로젝트는 다양한 사적 혹은 공적인 경관들에 걸쳐서 우리의 경험이 연장되고 통합되는 "사이-장소"

를 만들기 위한 시도다. 사이-장소는 사적인 삶과 공적인 삶을 동시에 가능하게 하는 "전이 공간"으로 작동하기 때문에 그 경계는 깊다. 그리고 그 깊은 경계는 사람들이 서로 만나고 소통하게끔 한다.

방음벽과 옹벽으로 반-공적공간(semi-public space)이 소거되고, 사유화된 발코니로 반-사적공간(semi-private space)이 닫혀지고 있는 전형적인 닫힌 경계에 대항하여, 우리는 깊이 있는 경계면 안에 참여와 공유를 촉진하는 "공적인 사사성" 혹은 "사적인 공공성"의 공간으로 기능하는 전이 공간(transitioanl space)을 디자인하길 의도하였다.

우리의 작업은 경계를 활성화하는 네 가지 다른 방식과 여러 가지 개연성 있는 공존의 방식을 보여준다. 간단히 정리해 보면, 여기서 제안된 디자인은 두 가지 방향성을 가지고 있다. 하나는 전이 공간을 개개인의 개별적인 참여와 집단적인 공유를 유도하는 반-사적공간으로 전환시키는 것이다. 그리고 다른 하나는 전이 공간으로부터 반-공적공간을 만들어냄으로써 집단적인 참여와 개개인의 개별적인 공유를 촉발하는 것이다. 이와 같은 실험적인 디자인전략은 공공성이란 개인의 사적 영역과 공적 영역 사이의 상호관계를 통해서 생성되는 것이라는 우리의 믿음을 반영한 것이며, 이 가운데 고착된 유형적 접근을 넘어서는 "사용의 위반"으로부터 새로운 경계의 영역이 생성될 때 열린 공공성이 확보된다는 것을 말하고자 함이다.

극단적인 사유화를 대신해서, 우리의 제안은, 사람들의 행위와 결합된 개별적 소공간들을 건축화하고 이들의 네트워크로 존재하는 공공의 경관을 조율함으로써, 미분화된 삶에 대응하여, "함께 존재하는 다양한 방식"을 보여주는 건축적 공간을 찾는 것이다. 따라서 이 공간은 개인적인(individual) 동시에 집단적인(collective) 공간을 가지며, 다른 무엇보다도 서로 다른 경관들(scapes)이 충돌하고 연결되는 전이의 공간(transitional space)을 갖는다.

결론적으로, 우리는 사적공간과 공적공간을 연결하는 반-사적공간이나 반-공적공간과 같은 전이공간을 디자인하고자 한다. 컴퓨터 스크린 위에서 키보드의 스페이스바가 작동하는 것과 마찬가지로, 전이공간은 개별적인 특성과 행위를 분절시키는 동시에 연계시키서 다양한 이야기와 경관들을 전개시킨다.

아파트로 대변되는 극단적인 동질성의 공간으로 양극화되고 단절된 한국적 상황에서는 개별적인 헤테로토피아적 행위와 경험들을 촉발하고 만나게 하는 인터페이스적 공간과 이러한 경험들을 도시적 공공의 경험으로 지속시키고 연결시키는 네트워크적 공간이 더더욱 절실하다. 공공성(公共性: publicness) 혹은 공공 공간이란 사사성(私事性: privateness) 혹은 사적 공간과 별개의 공간이 아니라, 언제나 이들 공간의 교섭과 내번을 통해서 존재하는 것이다. 이것은 모든 흐름이 차이를 통해서 발생하는 원리나, "모두의 공간이 누구의 공간도 아니다(everybody's space is nobody's space)"는 말이 의미하는 바와 통한다고 하겠다. 따라서 우리는 언제나 개별적인 공간(individual space)을 가장 컴팩트하고 효율적으로 디자인함으로써 그 가운데 변화의 가능성을 찾고, 새로운 집합 형상(collective form)에 대한 일종의 알리바이를 확보하여, 일반적인 형상으로부터 탈출하여 혁신적인 새로움을 유도해내고자 한다.

The Ways of Being Together

Activating boundary

"A public space is not just a given open or accessible space. Likewise, a private space is not just an enclosed space, but both spaces are being made between inter-spaces where individual experiences take with the other's."

Barricaded city

In the contemporary fragmented urban space such as gated community or Danji (typical apartment housing complex in Korea), a utopian dream, which is aiming for ideal community like a modern did, pervades again, facing with various heterotopian spaces initiated by individualization, urbanization and mobilization. However, people in this community are always meeting the same kinds of people, being demanded to do the same thing, feeling the same class consciousness and deleting other people°Øs existence as this community is actually an inter-oriented spaces filled with homogeneous value only, in most cases.

Even though they still seem to get together, meet and communicate, they live the fragmented or disengaged life within a barricaded community and fixed regular routines. Moreover, this barricaded space of homogeneous value disconnects a myriad of differential life and multi-layered places where these lives exist, thus blocks up creative possibility of which initiation takes place in the collision of multiple value and action.

In the cases of ready-made social class consciousness by living floor areas, fixed spatial value by brands of apartments, or undefeatable preference for fortified Danji with soundproof walls and retaining walls, we can find a segment of invisible but segregating walls in our society.

In fact, the apartment Danji, typical living space complex in Korea shows an extreme case of private-ness. In this space, both semi-private space and semi-public space are extremely reduced or deleted in order to get ultimate private space. Whereas, unbridled heterogeous desires and activities prevails outdoor space over the fence of housing complexes, but these are encapsulated and privatized in various types of rooms, symbols and kits, which internalize urban public life and separate us from others, as well.

Activating Boundary to make inter-places

The project we are proposing aims to make an "inter-place" where our experiences are extended and integrated across various private and public scapes. Its boundary is deep because it works as a "transitional space" that allows private and public life at the same time, and this deep boundary make people to meet and communicate with each other.

Against the typical enclosed boundary, where semi-public space is deleted through soundproof walls or retaining walls and where semi-private space is closed through privatized balcony, we intended to design a transitioanl space whitch acts as a space of "public privateness" or "private publicness" that promotes participation and sharing within its deep boundary.

Our works shows 4 different ways of activating boundary, and many possible ways of being together. In brief, the proposed design has two different directions. One is to transform transitional space into semi-private space which allows individual participation and collective sharing. while the other is to make semi-public space out of transitional space, which initiates collective participation and individual sharing. This experimental design strategy reflects our belief that publicness could be generated from an inter-relationship between private and public domains, and its openness might be acquired in the process of interaction, only if new boundarys were to be produced from a 'violence of use' beyond a fixed typological approach.

In place of the extreme privatization, the proposal is to seek an architectural space that shows "the various ways of being together" in response to our differential life, by architecturing individual spaces which are strongly connected with various users' activities, and by mediating public scape which exists as a network of them. Thus this space has both individual and collective space at the same time, and more than anything else, transitional space where different scapes are colliding and being connected with each other.

In conclusion, we attempt to design a transitional space, such as semi-private space and semi-public space, which connects private space with public space.
Like the spacebar on a keyboard does on a computer screen, it articulates individual characters and actions, and links them simultaneously to develop various scenarios or scapes .

We believe that *the space of interface*, which prompts and contacts heterotopian individual activities and experiences, *and the space of network*, which connects and links them into urban public experience, are highly required in the case of Korea, where spaces of cityscape are polarized and disconnected in the space of extreme homogeneity represented by apartment housing, and that Publicness or public space is not separate from privateness or private space as it exists by constantly interacting and facing with these spaces. Its principle is that every flow is generated from the differences. However, it is somehow the same as the phrase, 'everybody's space is nobody's space'.
Therefore, we try to design our individual space more precisely and effectively, in approach to find possibilities of change among spaces and secure a kind of alibi about a new collective form. By escaping from a general form, we can bring about innovating freshness.

vertical boundary — 3P house / House Crossover

horizontal boundary — Extruded Community / House Cornerstone

diagonal boundary — Street Pool Villa / Street Housing

virtual boundary — Public Garden House

activating boundary to make transitional spaces
to make inter-places

operational engravings for communication

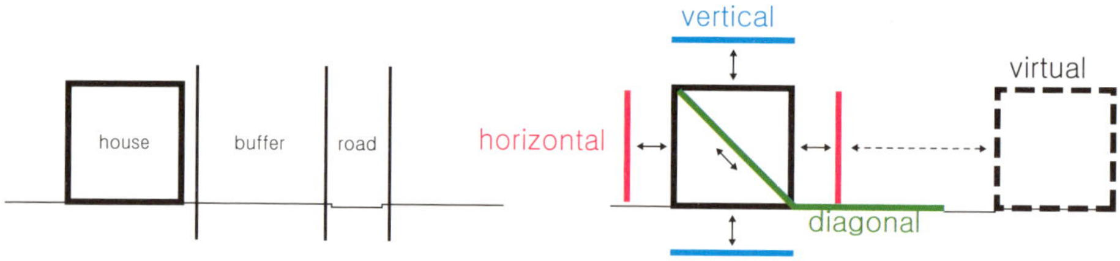

fragmented urban parcel
buffer-seperator

interlocking urban boundary
interface-mediator

106
105
LOTTE CASTLE
107

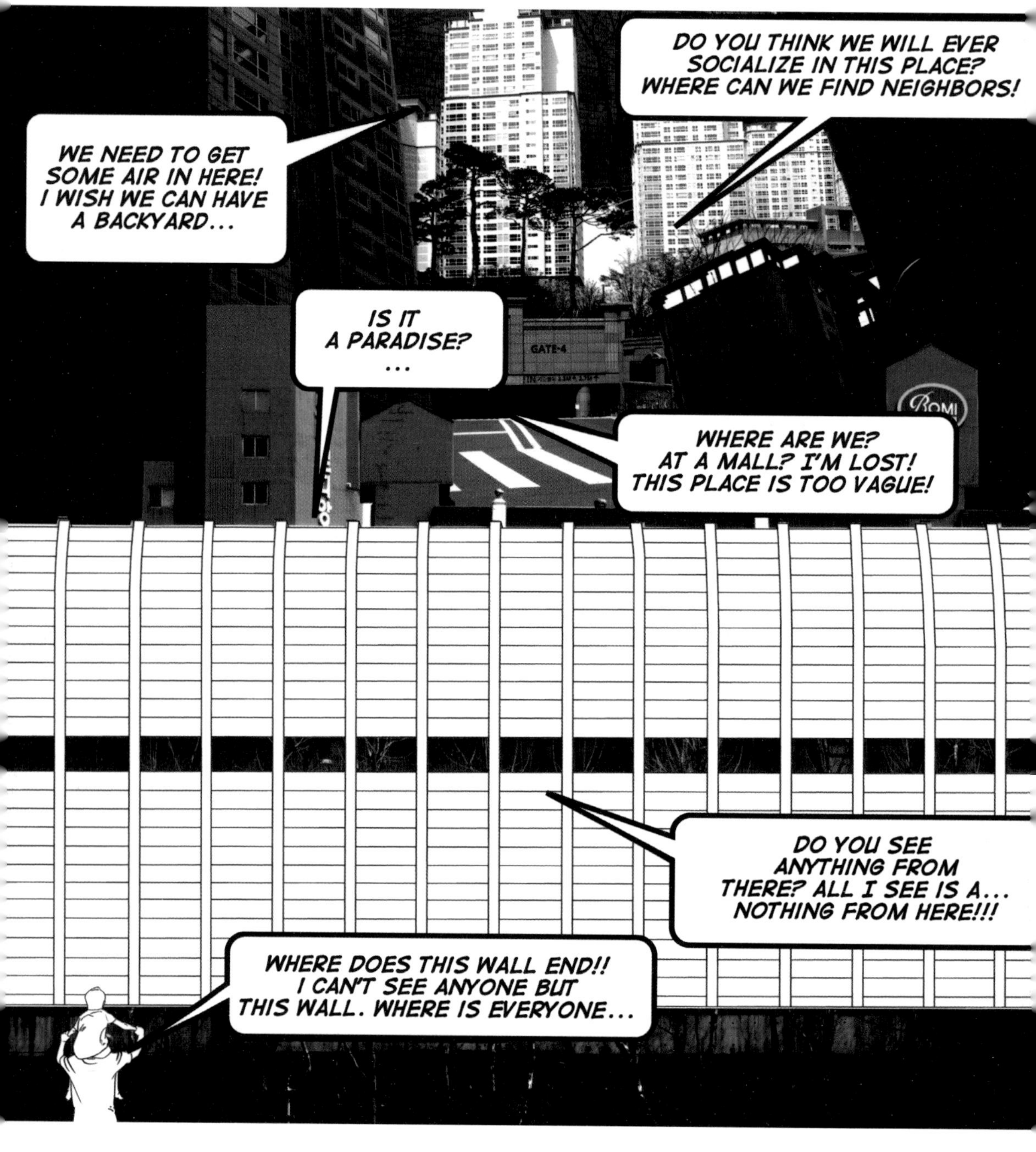
WE NEED TO GET SOME AIR IN HERE! I WISH WE CAN HAVE A BACKYARD...
DO YOU THINK WE WILL EVER SOCIALIZE IN THIS PLACE? WHERE CAN WE FIND NEIGHBORS!
IS IT A PARADISE? ...
GATE-4
WHERE ARE WE? AT A MALL? I'M LOST! THIS PLACE IS TOO VAGUE!
DO YOU SEE ANYTHING FROM THERE? ALL I SEE IS A... NOTHING FROM HERE!!!
WHERE DOES THIS WALL END!! I CAN'T SEE ANYONE BUT THIS WALL. WHERE IS EVERYONE...

IT IS REALLY HARD TO WALK OUTSIDE... THERE ARE A FULL OF THERE ARE FULL OF SYMBOLS, STANDING SIGNBOARD, BILLBOARDS, OBSTACLES,...
THERE IS NO VOID!!! ...
GIMME CIGARETTE

CONTENTS

ACTIVATING VERTICAL BOUNDARY

3P House

봉천동 다세대주택

Bongcheon Multi-family Housing

주어진 조건과 선택한 아이디어

일반적으로 다세대주택은 전용율이 대략 60%선으로 대개 법적으로 제한된 4개층의 주거공간, 1층의 공용홀과 주차장, 지하 근생시설로 유형화 됐다. 또한 주차 공간확보를 위해 전면도로에 접해서 주차장을 일렬로 배치하고, 최대 건폐율과 용적률 확보를 위해 인접 건물과 법이 허용하는 최소 이격거리를 사이에 두고 다닥다닥 붙여서 배치된다. 문제는 이러한 공간구성방식이 거리를 사이에 두고 인접해있는 도시 공간의 켜를 가로막기 일수고, 늘 경험하는 바와 같은 답답하고 꽉 막힌 거리공간을 만들어낸다는데 있다. 실제로 해당 사이트를 점유하고 있던 이전 건물 역시 후면의 막다른 골목과 완전히 차단되어 있었기 때문에, 막다른 골목에 인접한 거주자들은 항상 어둡고 좁은 골목을 두고 주차전쟁을 치르고, 비좁은 골목으로 통행해야 하는 불편을 겪고 있었다. 우리는 이전의 주택 프로젝트와 마찬가지로, 봉천동 다세대 주택 역시 디자인 초기단계에서부터 실용적이면서도 차별화된 공간구성방식을 시도하고자 했다.

The given condition and the selected ideas

In general, the ratio of exclusive private area for a multifamily housing units is about 60 percent and legally categorized into four stories of residential space and each story for common hall, parking lot and underground neighborhood commercial facilities. Also, to have parking space, parking lots are arranged in parallel, bordering on the front street. Likewise, to take advantage of the maximum building coverage ratio and floor area ratio, they were placed adjacent to each other with the legally allowed minimum clearance between them.

The problem is that this space composition frequently blocks much of the layer of adjacent urban spaces. As we often witness, it creates a heavy-looking and blocked up street space atmosphere. In fact, the previous building that occupied the relevant site was also completely blocked from the dead-end back street. Thus, the residents living adjacent to the dead-end street had to constantly bicker when parking their cars in the shady backstreet. Moreover, they also had difficulties transiting through a narrow backstreet everyday.

3P House

봉천동 다세대주택은 Public Pathway, Private Courtyard 그리고 Public Courtyard 이렇게 3개의 P가 있답니다.
옥상마당에 오르면 우리 동네가 한눈에 펼쳐져요! 항상 이렇게 편하게 이용할 수 있는 마당이 있어서 너무 좋아요!
이집 주인댁에는 마당이 세개나 있대요!!! 어떻게 한 건물에 마당이 세개나 있을까요??
옥상에 심었던 화분은 잘자라고 있나?
건축주는 최상층에 한옥 중정같은 마당을 원했지만, 5층전체를 주인세대에 할애하는 대신, 남측 4,5층을 점유하는 복층형식으로 주인세대를 계획하였대.

IDEOGRAM

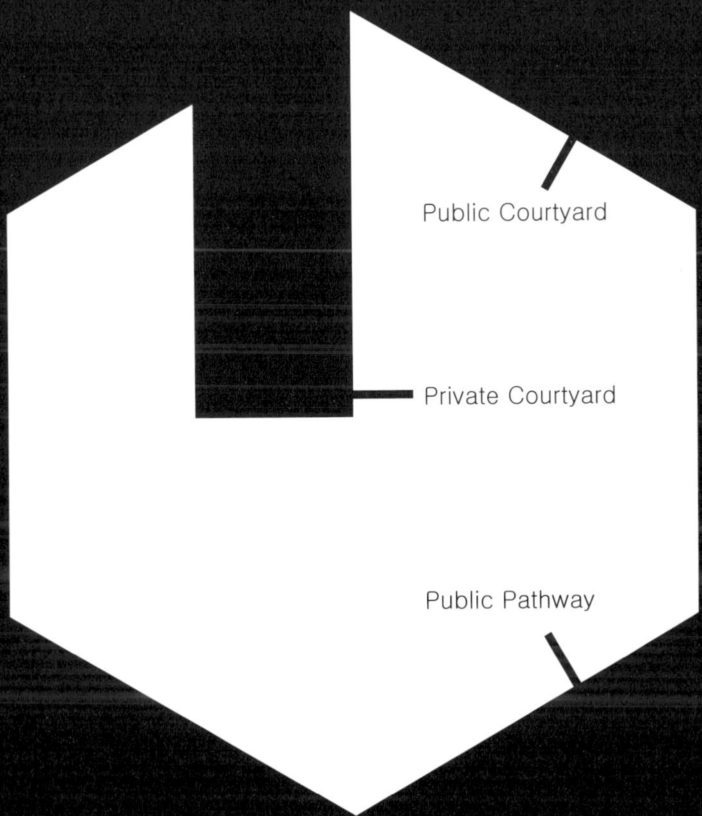

DIAGRAM

Public Pathway Private Courtyard Public Courtyard

공공성의 확보와 공간의 자유도를 높이기
securing greater publicness and increasing the degree of freed up space

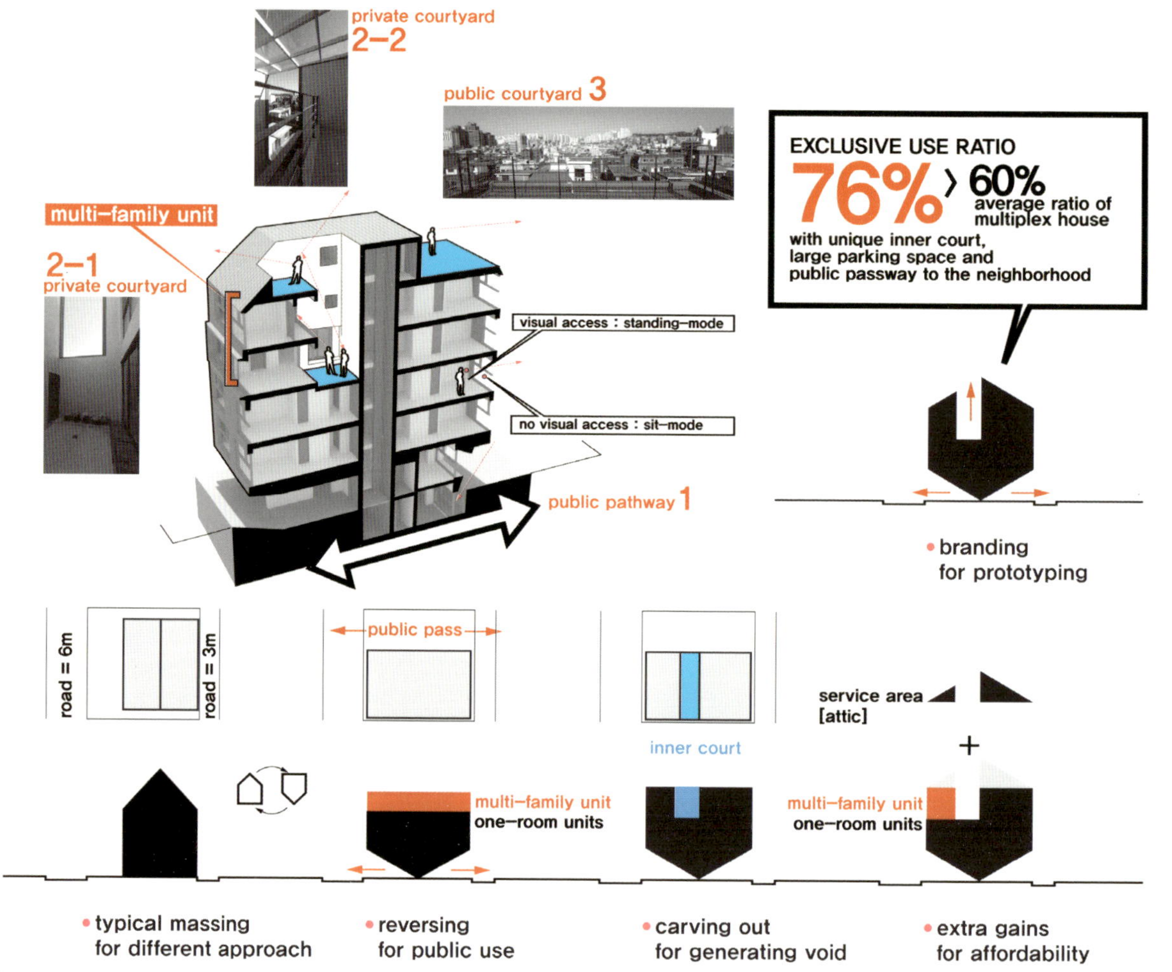

DRAWINGS

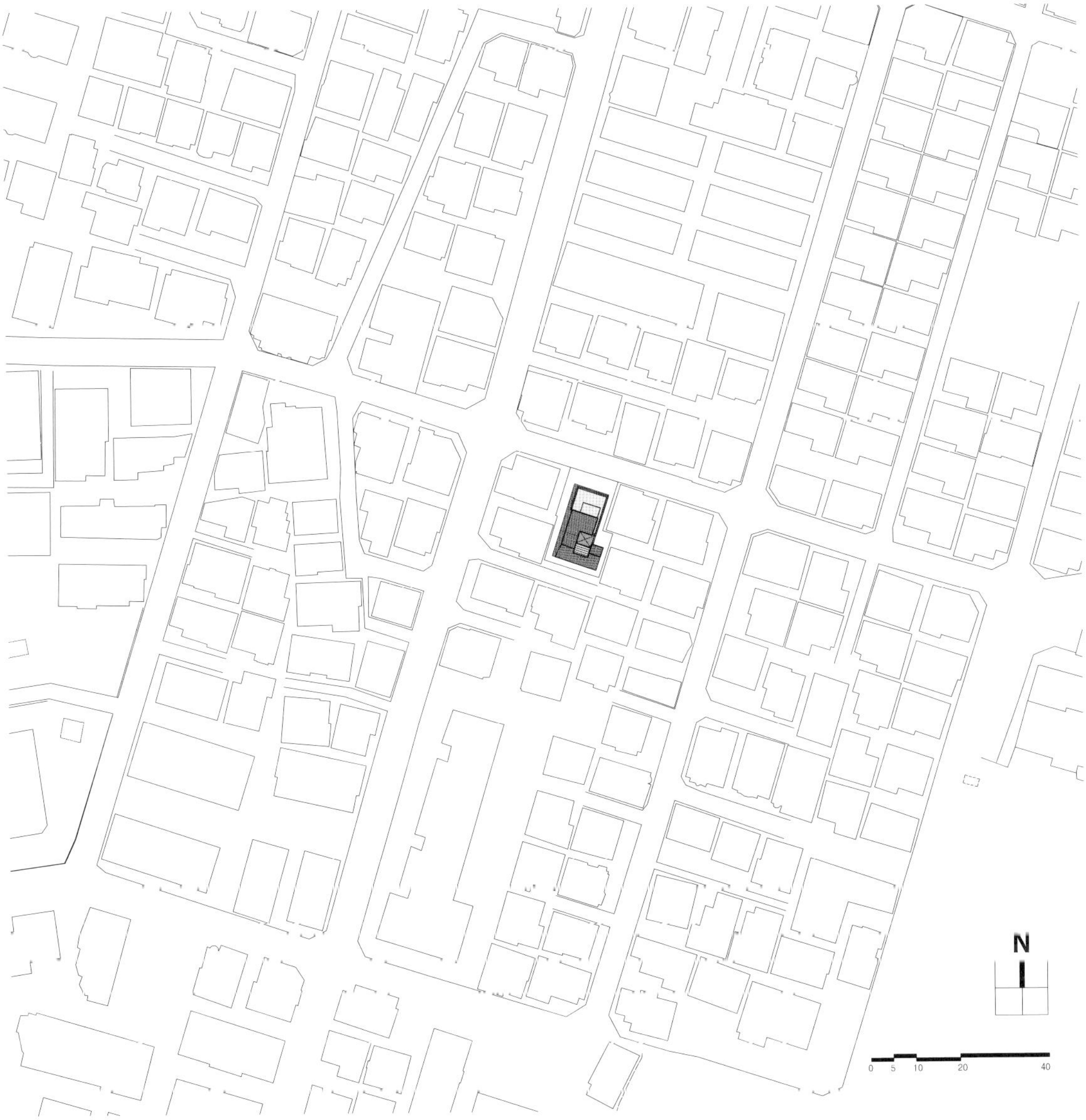

DRAWINGS

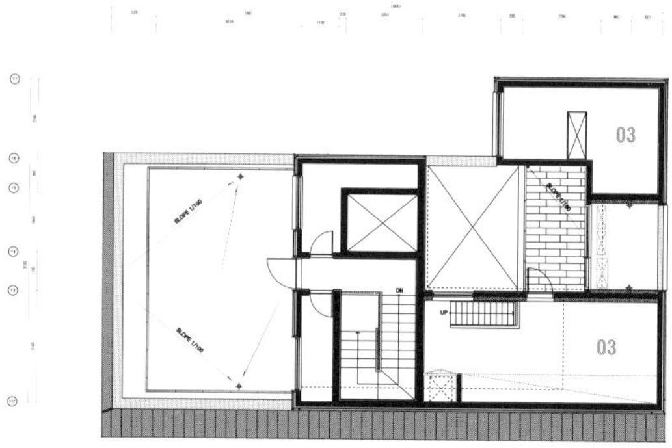

garret floor plan

fifth floor plan

fourth floor plan

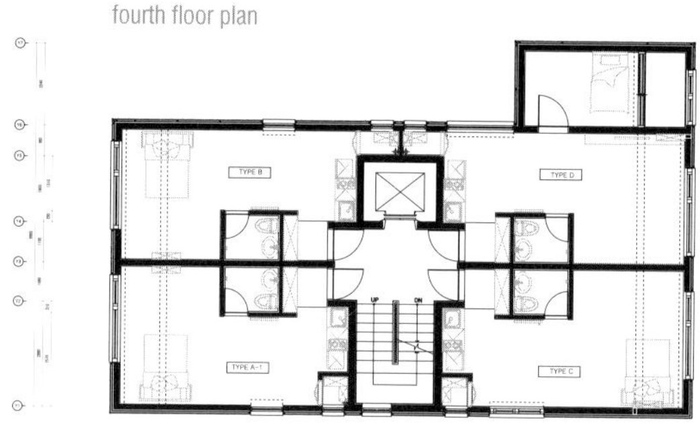

second, third floor plan

01_kitchen
02_bedroom
03_garret
04_dressroom
05_bathroom

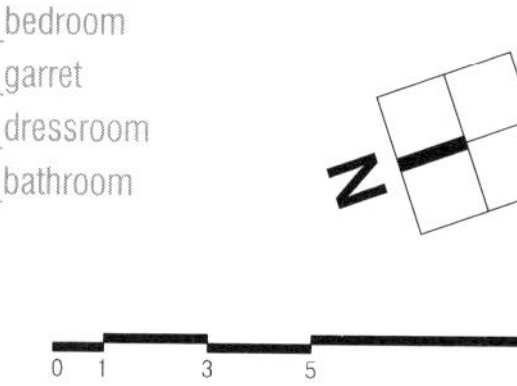

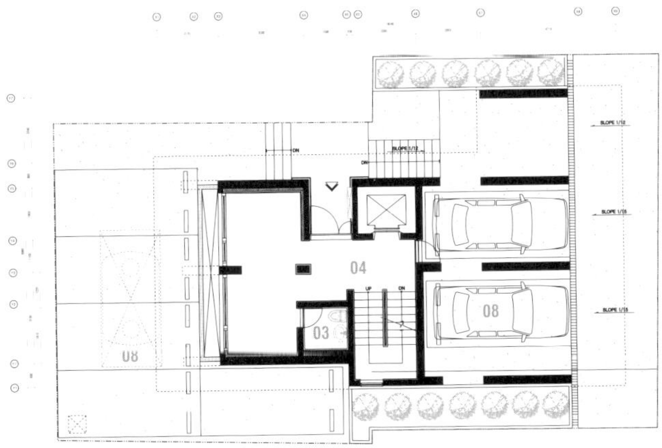

01_bedroom
02_kitchen
03_bathroom
04_hall
05_livingroom
06_elev. hall
07_garret
08_parking lot
09_storage

first floor plan

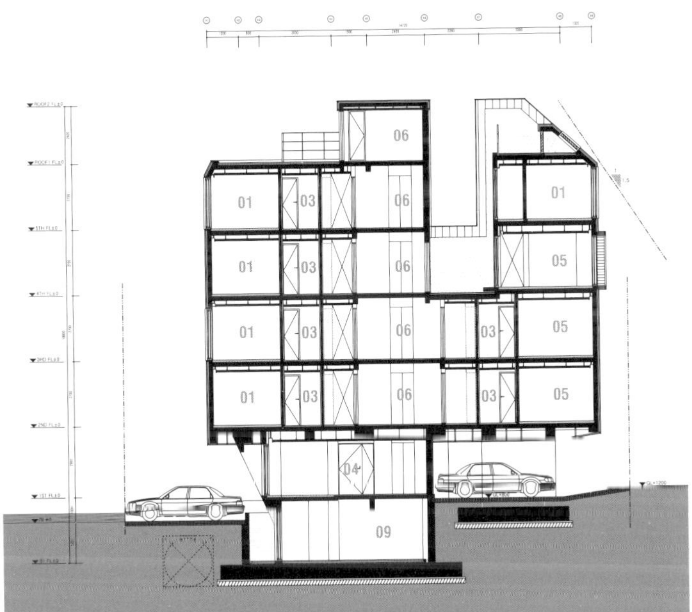

section_01

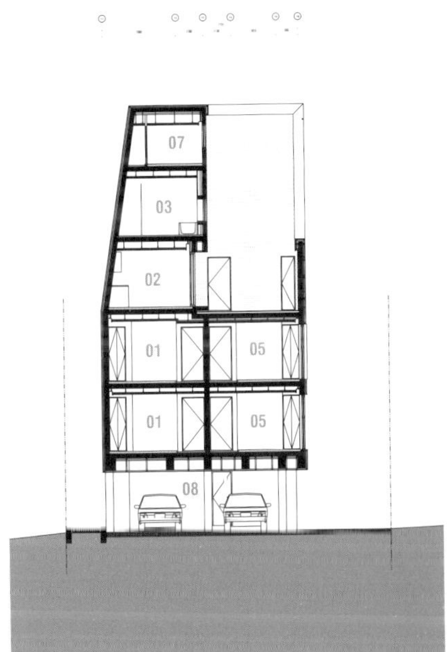

section_02

0 1 3 5

공용공간을 공공공간으로 전환시키기

대지와 법규 분석을 통해 최대 건폐율과 최대 용적률을 확보하기 위해서 반드시 전면도로에 면해서 길게 배치할 필요는 없다는 점을 발견했고, 치밀하게 공간의 폭을 줄여 외부에서는 건물의 전후방 거리를 연결시키는 공공로(public pass)를 확보했다. 내부로는 복도공간을 줄여 불필요한 공용공간(common space)을 제거하는 다이어트를 감행했다. 요컨대 공용공간을 공공공간으로 변화시켰다.

Converting common space into public space

Through an analysis on land and regulation, we found that it is not necessary to construct the building along the front street, in order to take full advantage of the maximum floor area ratio. Also, a public pass was allowed that connects both the back and front street of the building by precisely minimizing the width of space.

Internally, the corridor space was slimmed down as a way to remove some of the common space area. That way, common space was converted to public space.

공용공간을 전용공간으로 전환시키기

13세대에 요구되는 주차공간 확보를 위해 전형적인 집의 형상을 거꾸로 뒤집었다. 1층은 근생시설이, 2층부터 5층까지 주거공간이 놓인다. 뒤집힌 형상을 위해 공용 홀 공간을 최소화하고 내부로 기울어진 구조기둥과 함께, 엘리베이터 홀을 중심으로 십자형을 이루는 구조벽이 하중의 대부분을 받는 구조 시스템을 채택했다.

또한 이 십자형 구조벽에 계단실, 화장실, 보일러실, 신발장, 덕트 등의 유틸리티 공간을 최소 유효치수를 기준으로 치밀하게 집중시켜, 최대 전용면적을 확보했다. 반복된 최적화를 위한 설계 과정 중 공용공간 일부가 전용공간(exclusive private space)으로 탈바꿈되었고, 이러한 과정을 통해서 전용율(exclusive use ratio) 76%에 다다랐다. 이러한 접근은 공간의 자유도를 높이기 위한 우리의 일관된 관심사이기도 하다. 한편 최소 유효치수의 상당 부분은 자료집성에 기록되지 않은 다세대 건축 시장에서만 사용하는 치수였다.

Converting common space into private space

To obtain parking space required for 13 households, a typical house shape was turned upside down. The first floor was set aside for community facilities. Residential space occupies the second to fifth floors. For the converted shape, the common hall space was minimized and we chose a structural system such that the majority of building load was placed on the structure wall that forms a crossed-shape around the elevator hall, along with the structural column slanted towards the inside. The maximum exclusive private space was obtained by designing the utility spaces as close as possible to the cross-shaped structure wall. These included a stair room, washroom, boiler room, shoe room, duct area, etc. that was designed based on the critical dimension (minimum dimension). During the design process to achieve repetitive optimization, a part of the common space area was converted into the exclusive private space, resulting in reaching up to 76 percent of exclusive use ratio. This approach has always been our main interest to increase the openness of overall space. On the other hand, the majority of the critical dimension (minimum dimension) is the size that is used only in multifamily construction market units.

5층 아이들방에서
다락으로 연결되는 공간이에요.
정말 알뜰하게 공간을 활용해서
사용하고 계신답니다. 주인 가족분들이
책이 많아서 고민이셨는데, 곳곳에
마련해둔 수납공간을 잘 사용하고
계신가 봅니다.

Public Pathway / Private Courtyard / Public Courtyard

건축주는 최상층에 한옥 중정과 같은 마당을 갖기를 원했지만 최상층인 5층 전체를 주인 세대에 할애하는 대신, 십자형 구조벽과 코어로 인한 남북으로 양분된 볼륨의 특성을 활용해 남측 4,5층을 점유하는 듀플렉스(duplex) 형식으로 주인 세대를 계획했다. 거실에 면한 2개 층 높이의 깊이 있는 중정, 다락층에 면한 옥상 마당, 봉천동 일대의 파노라마를 담는 북측의 옥상 마당 등 총 세 개의 마당을 계획했다. 특히 북측 옥상 마당은 코어를 통해 다른 임대 세대가 자유롭게 접근 할 수 있어, 최상층 주인세대가 옥상을 전용하는 일반적인 다세대와는 다른 공공적인 성격이 부여됐다.

Public Pathway/Private Courtyard/Public Courtyard

The owner wanted to have a courtyard on the rooftop as is initiated in traditional Korean house. Instead, by making use of the volume divided from north to south with a cross shaped structural wall and the core, it was designed as a duplex style space where the owner occupies both the entire fifth floor and the fourth and fifth floor part of building in the south.

A total of three courtyards were designed including the courtyard that stretches the length of two stories in the living room, another courtyard situated on the rooftop that is in the loft and the last one on the rooftop where one can have a panoramic view of the Bongchun-dong area. In particular, the rooftop courtyard in the north can be accessed by the tenants. This appends a more public aspect to the project compared with other multifamily housing units where the owner strictly prohibits others from using the rooftop space

다세대 주택에 대한 생각

이 프로젝트는 공공성의 확보와 공간의 자유도를 높인 점 등 우리가 추구하는 건축적 접근 방식을 적절히 대응시킨 작업이라고 생각한다. 프로젝트 조건의 최소와 최대 값을 경험하는 것이 우리 일이라면 다세대 주택은 최소 조건의 일이 아닐까? 사실 개별 프로젝트의 한계 안에서 건축가의 능동적인 선택을 통해 전체 다세대 주택 시장에 미칠 영향력은 미미할 것이다. 현재 다세대 주택은 도시의 섬 혹은 블랙박스다. 따라서 블록 단위 개발이 필요하다고 본다. 법개정을 통해서 공동 개발을 할 경우 지하 혹은 지상 주차장을 공동으로 사용하거나 합벽 건축을 가능하게 한다면 현행 용적율 한계내에서도 추가면적이나 오픈스페이스를 확보할 수 있는 여지가 생길것이다. 결국 공공성이나 사사성을 동시에 고려하는 선택과 집중의 전략이 있어야 저층형 고밀도 도시 공동주택으로서 다세대 주택의 가능성이 발견될 수 있을 것이다.

여지를 디자인하기

봉천동 다세대 주택은 디자인의 여지(affordability)를 통해서 공공성에 접근하고자 한 프로젝트다. 이는 엇갈린 듀플렉스(skipped duplex) 형식의 단위세대를 적층해 4개 층의 전 세대가 지하와 옥상에 서비스면적을 갖도록 한 남양주 타운하우스 프로젝트(2007년), 듀플렉스와 스킵플로어(skip floor) 형식을 결합한 단위세대를 연이어 병치(juxtaposition)해 복도와 같은 공용공간 대신에 전 세대를 연결하는 공공 정원을 만들고자 했던 베니스 비엔날레 출품 계획안(2010년), 치밀하면서도 공간 활용도가 우수한 공장생산 모듈러주택 프로토타입 프로젝트인 스카이홈(2009~, SK D&D와 진행) 등에서 실험해온 일관된 공간 디자인 전략 가운데 하나다.

공공성(公共性: publicness) 혹은 공공 공간이란 사사성(私事性: privateness) 혹은 사적 공간과 별개가 아니라, 언제나 이들 공간의 교섭과 대면을 통해서 존재한다. 이는 모든 흐름이 차이를 통해서 발생하는 원리나, '모두의 공간이 누구의 공간도 아니다(everybody's space is nobody's space)'는 말이 의미하는 바와 통한다고 본다. 따라서 우리는 언제나 개별적인 공간(individual space)을 가장 치밀하고 효율적으로 디자인하고자 하며, 그 가운데서 변화의 가능성을 찾고 새로운 집합 형상(collective form)에 대한 일종의 알리바이를 확보한다. 그렇게 일반적인 형상으로부터 탈출하여 혁신적인 새로움을 유도한다.

Thoughts on multifamily housing

we believe that this project has been made based on the architectural approach that we pursued, such as securing greater publicness and increasing the degree of freed up space, etc. If our task is to experience the minimum and the maximum of project requirements, then I believe that a multifamily housing project is a project with the minimum requirements.

In fact, with the restrictions imposed on each project, it would be almost none that a proactive choice of an architect may have an influence on the entire multifamily residential housing market. Currently, the multifamily housing market is as small as an island or a black box in the city. Therefore, it is necessary to develop it on a larger city block scale. If it is possible to conduct joint development with the help of revisions to the relevant regulations, there would be room for an open space or additional area even with the current restriction on the floor area ratio, once either underground or in a ground car park which can be jointly shared or it is okay to build a shared wall. There is a possibility in the future for multifamily housing units as low story apartment housing units in a city with high population density, only if there is a strategy to choose simultaneously and focusing on public-ness and private-ness.

Design Affordability

This multifamily housing project in Bongchun-dong is a project that deals with publicness through design affordability. This is one of our design strategies, consistently carried out on many projects. For instance, a townhouse project in Namyangju (2007) was planned in proposal of individual units that were stacked four stories high in a skipped duplex style, where every household has a service area in both underground and the rooftop. Another example is a design plan submitted to a Biennale in Venice (2010) that household units are placed side by side (juxtaposed) by combining a duplex with a skip floor type, such that a public garden was constructed to connect every household. A prototype project, so called 'SKY Home' (2009 ~, collaboration with SK D&D) was another example with a factory-produced modular house, which is excellent for making better use of a space.

Publicness or public space is not separate from privateness or private space. It exists by constantly interacting and facing with these spaces. Its principle is that every flow is generated from the differences. However, it is somehow the same as the phrase, 'everybody's space is nobody's space'. Therefore, we try to design our individual space more precisely and effectively, in approach to find possibilities of change among spaces and secure a kind of alibi about a new collective form. By escaping from a general form, we can bring about innovating freshness.

CHEAP AND DURABLE FINISHING MATERIAL
TRANSLUCENT WINDOW FOR PRIVACY, DESIGNED IN RESPONSE TO SEDENTARY LIFESTYLE
PUBLIC PATHWAY
46

HOUSEHOLD GOODS SEEN THROUGH THE WINDOW
SHOW EACH RESIDENT'S SPECIFIC LIFESTYLE
TRANSPARENT WINDOW FOR VENTILATION AND LIGHTNING
ENOUGH SPACE FOR PARKING

DESIGN

ACTIVATING VERTICAL BOUNDARY

House Crossover

남양주 타운하우스

Namyangju Townhouse

최대 용적율 안에서의 최대 가능성

이 프로젝트는 남양주 호평동 일원에 위치하는 공동주택 프로젝트로, 비록 시행사의 여건으로 인해 실제로 지어지지는 못했지만 우리가 추구해온 디자인 접근법이 가장 구체적으로 구현된 프로젝트 가운데 하나다.

건물이 위치하게 될 지역은 자연녹지지역으로 건폐율 20%, 용적률 100%, 지상 4층 이하로 건축물의 규모가 제약을 받는 곳이다. 4층 이하로 건축물의 층수가 규정되어 있기 때문에 용적률 100%는 불가능한 의미 없는 숫자로, 결국은 건폐율 20%, 용적률 80%를 꽉 채운 4층 건물이 이 땅에서 건축할 수 있는 최대 볼륨이었다.

이 프로젝트의 모순은, 의미 없는 용적률 100% 규정과도 같은 시행사의 요구조건으로, 용적률 최대를 요구하면서도 새로운 유형의 주거공간을 실험해보고 싶다는 것이었다. 일견 터무니없는 조건이기는 했지만, 우리 자신을 영점(zero)으로 되돌려 우리가 알고 있는 주택의 유형을 다시 점검하고 분석하는 계기로 삼기로 결정하고, 다양한 공동 주택의 유형별 분석에 들어갔다.

각각의 유형별 장단점과 함께 공간 요소들(core, master bedroom, utility spaces, etc.)을 재조합하고 재위치 시키는 작업을 통해서 유형상으로는 Duplex와 Skip Floor 사이에 위치하는 새로운 공간유형이 기존의 주택건설방식과 구조시스템을 사용하면서도 독창적인 새로운 단위주호(unit)를 만들 수 있음을 증명할 수 있었고, 이 유형의 3차원적 조합을 통해 4층 전세대가 다락방 혹은 지하실 형태의 서비스 면적(분양면적에서는 제외됨)을 확보할 수 있게끔 계획할 수 있었다.

더 나아가 우리는 이 새로운 유형이 공간적인 새로움과 시용공간의 확장뿐만이 아니라 설비의 집중화나 에너지효율의 증대 및 프라이버시의 확보와 같은 여러가지 가능성과 관계되어 있다는 사실을 발견해가며 그 특성을 적극적으로 드러내는 방향으로 디자인을 발전시켰다.

Maximum Possibility within Maximum FAR

This project is to build apartment houses in suburban area of Hopyung-dong, Namyangju-si. Though the plan could not be realized owing to some critical problems of the developer, it was one of the projects that our design approach was materialized in the most concrete way.

The building was to be constructed in a natural green area where buildable area is restricted under 20% of the site area and maximum FAR(Floor Area Ratio) is 100% of the site area. In addition, its height must be smaller than four stories by regulation. As it was restricted to build a building smaller than a four ground-story one, actually 100% of FAR is meaningless in this plan. In the end, 20% for the building footprint and 80% of FAR was the optimal maximum choice we could take.

However, the contradictory side of this project is that the developer demanded maximum freedom of imagination while they push us to take only maximum volume: it looks like strange capacity of 100% FAR by regulation which doesn't make sense at all. Anyway, they asked us to explore a new type of residential space within maximum volume and we decided to take it as an innovative opportunity to examine and analyze various types of house and its possibility from the zero point as if we do not have any preconception about it.

By assembling and relocating spatial elements such as core, master bedroom and utility spaces, we could prove that we can make a new and unique housing unit, which is located between "Duplex" and "Skip floor" in housing typology, while making use of the current typical construction technique and structural system. Through three dimensional combinations, we could make a plan with added service area, which is not counted as a rental area in Korea, such as a garret or a basement room to every apartment unit, for all four households.

Furthermore, we discovered that this new type of space is quite practical not only in terms of spatial flexibility and compactness, but also in energy and mechanical efficiency. Likewise, we developed the design, making the best use of the merits of the house.

HOUSE CROSSOVER

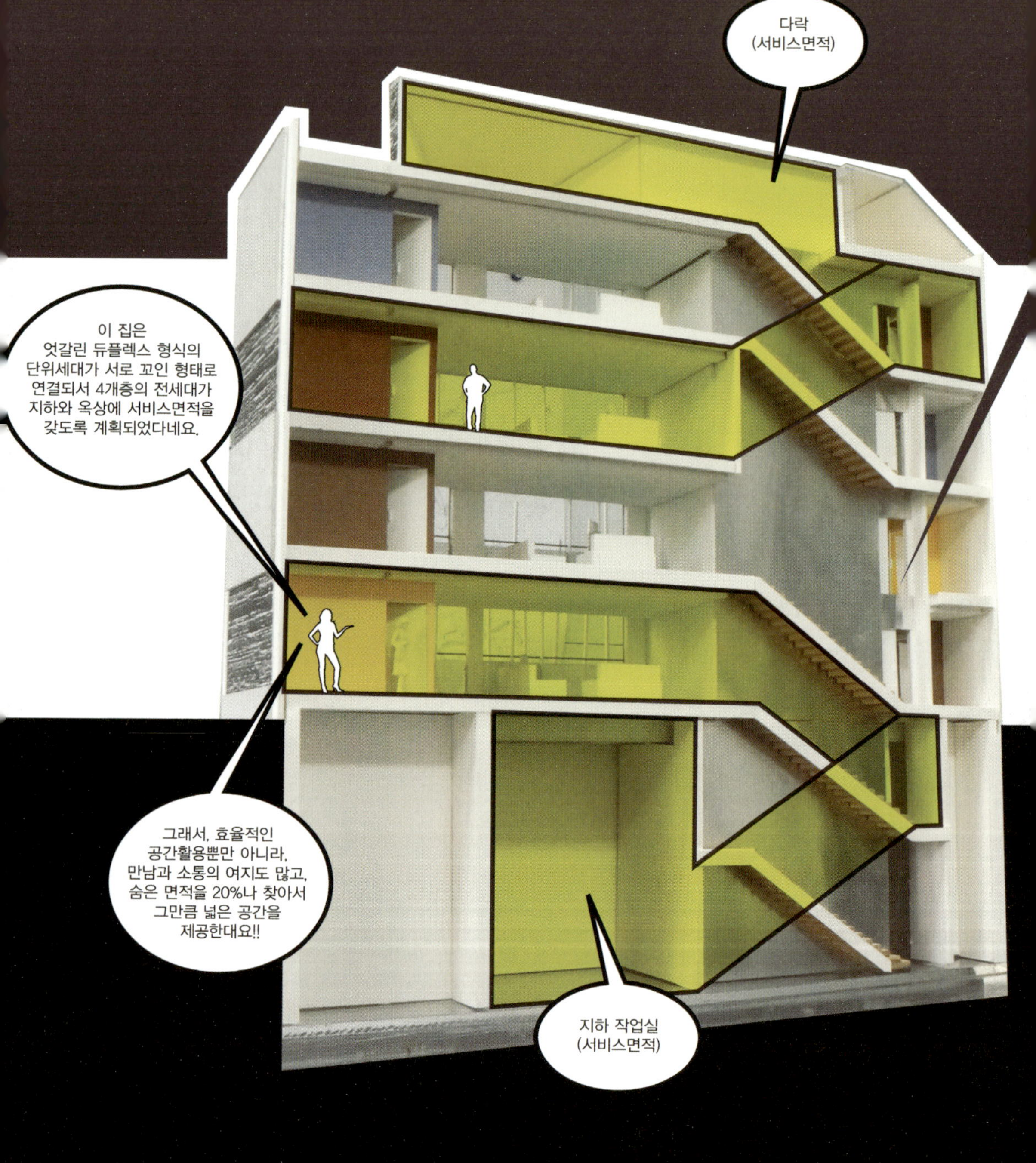

IDEOGRAM

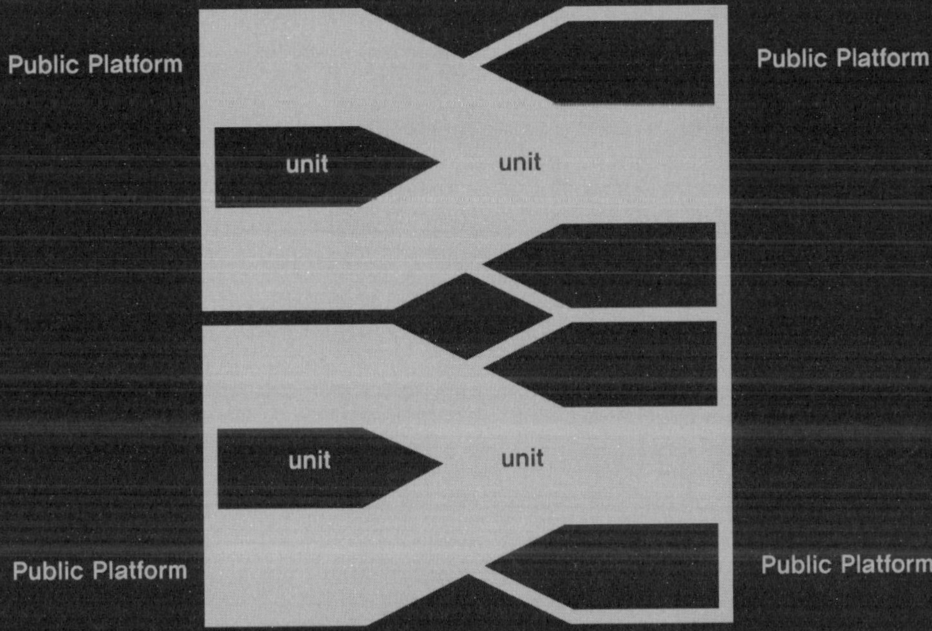
Public Platform
Public Platform
unit
unit
unit
unit
Public Platform
Public Platform

DIAGRAM

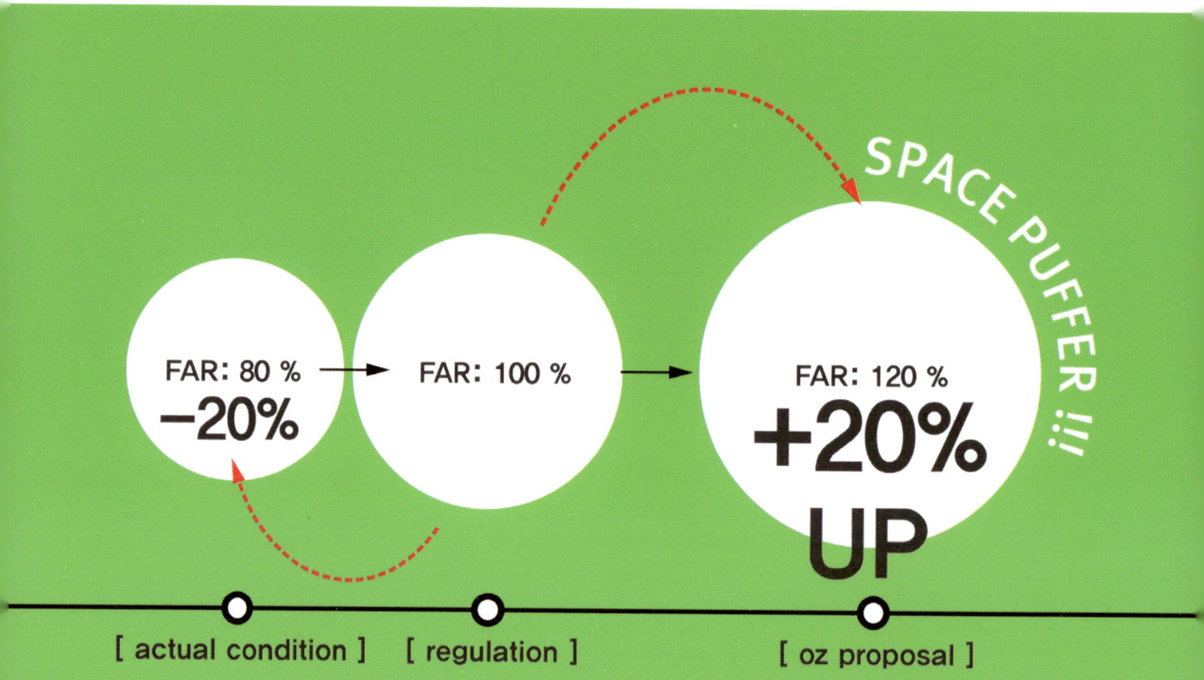
SPACE PUFFER !!!
FAR: 80 %
-20%
FAR: 100 %
FAR: 120 %
+20%
UP
[actual condition]
[regulation]
[oz proposal]

DRAWINGS

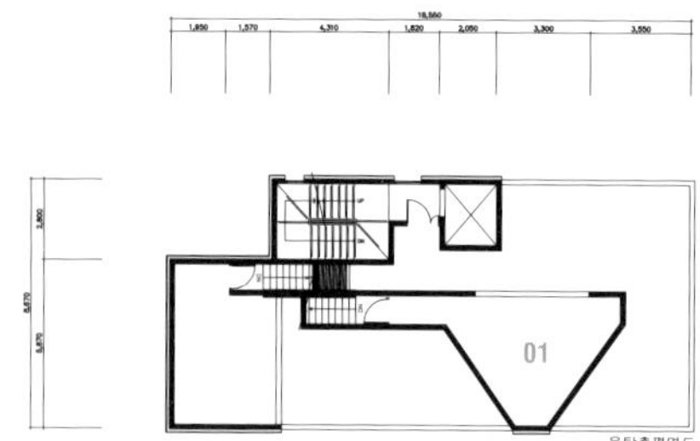

옥탑층평면도

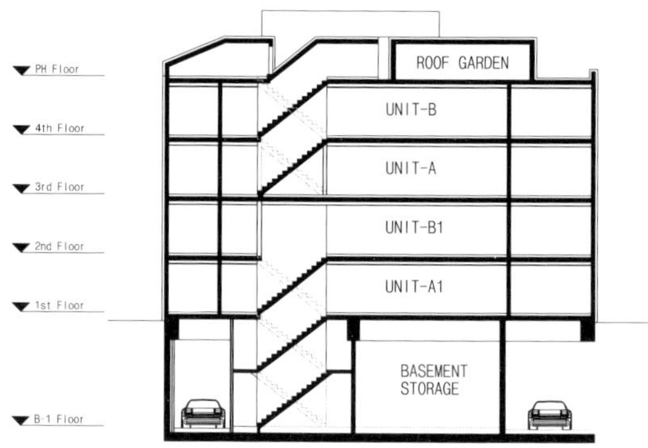

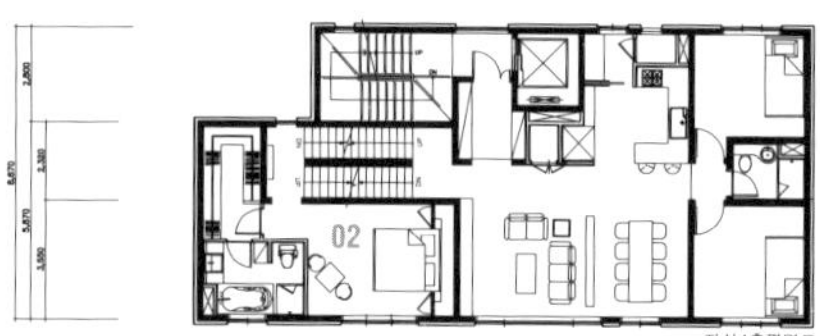

지상4층평면도

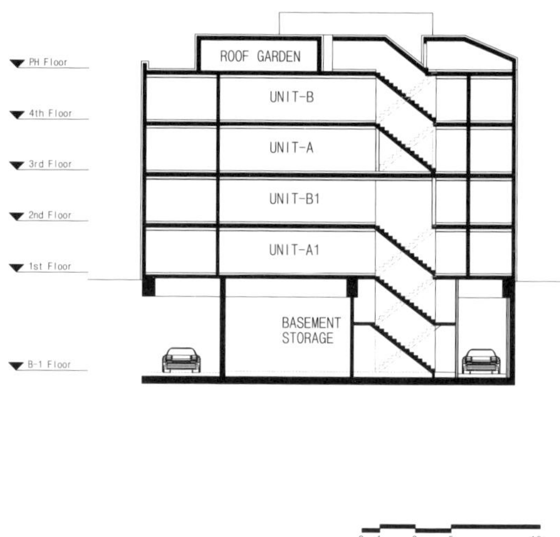

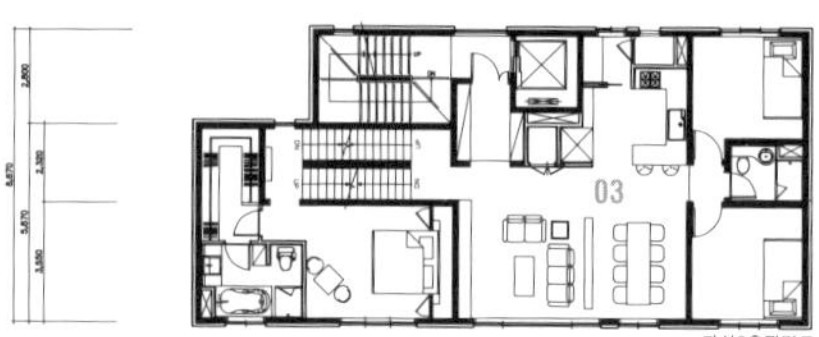

지상3층평면도

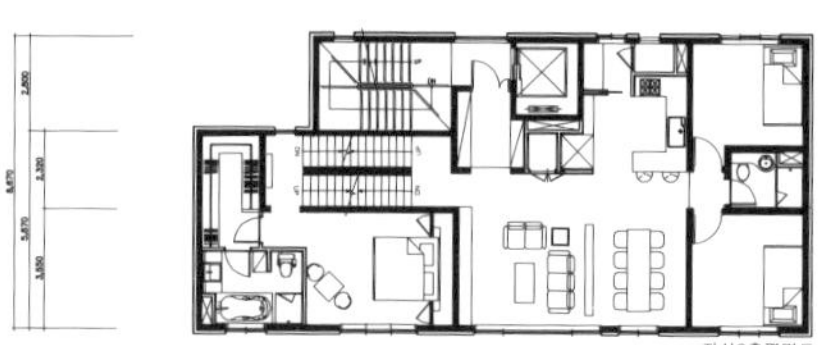

지상2층평면도

01_attic
02_master bed room
03_living room
04_basement storage

지상1층평면도

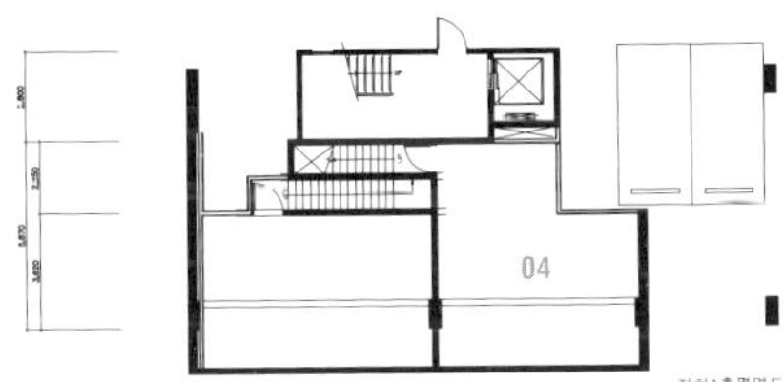

지하1층평면도

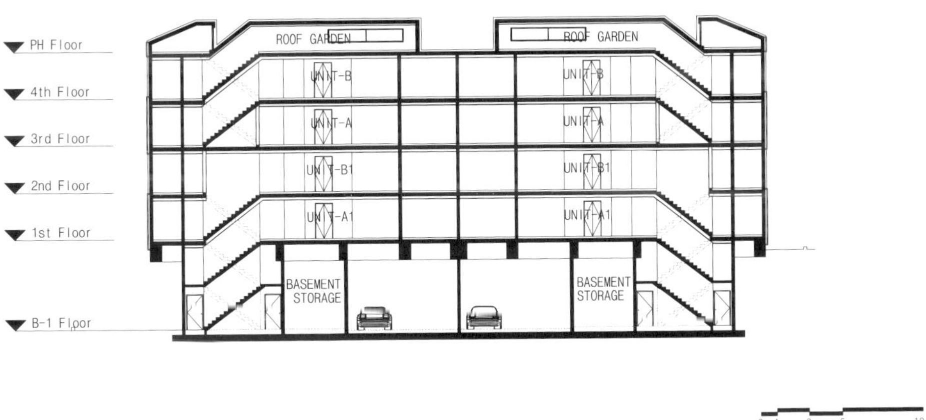

결과적으로 우리는 법정최대 용적률을 확보하면서도 5 Bay 이상의 최대 전망을 갖는 3개층의 단위주호로 구성된 4개층 공동주택을 설계할 수 있었다. 단위주호의 합으로만 보면 12개 층의 공간(4층*3층)이 압축되어 있는 셈이고, 주거영역의 연면적 기준으로 생각하면 용적률 120%가 달성된 셈이다. 용적률 40%에 해당하는 연면적증가와 독창적인 내외부공간의 확보는 주어진 문제를 뛰어넘어 복합적이고 체계적으로 문제의 공간을 재-조직하고 재-조합함으로써 얻어진 디자인 프리미엄(design premium), 즉 "디자인의 여지"(design affordability)임에 틀림없다.

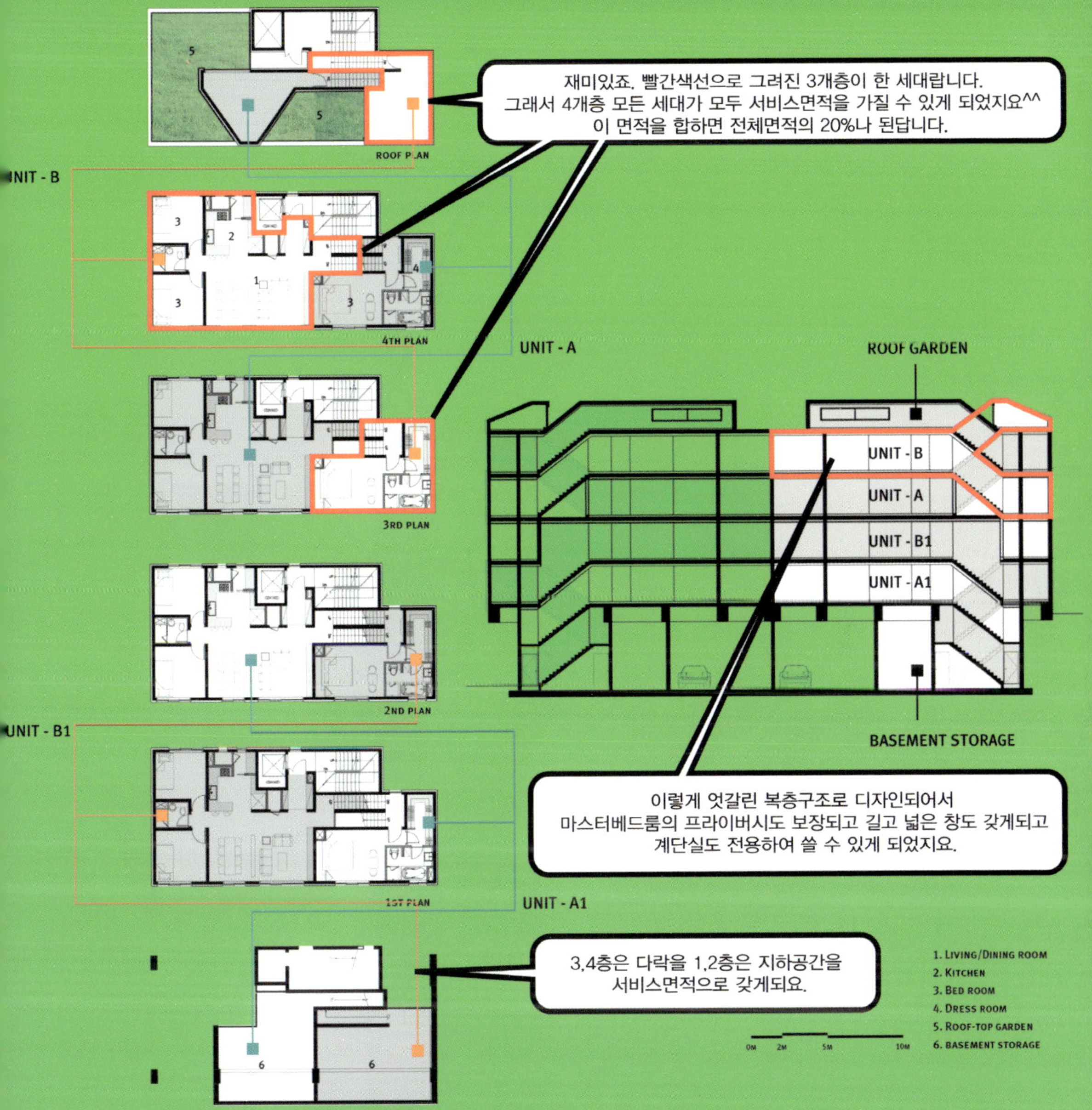

As a result, we come up with a four-story apartment house that is composed of 4 units of 3 different floors with a five-bay panoramic view.
Judging from the sum of the units, it is regarded as the building composed of 12 floors (4 units X 3 floors) and it can be interpreted that we succeeded to build a house which has 120% of Floor Area Ratio in the end. We believe the increased area of 40% FAR and unique interior and exterior spaces are surely a design premium or let say "design affordability" which has been achieved by re-organizing and re-assembling the spaces systematically.

요컨대, 이 프로젝트는 엇갈린 듀플렉스(skipped duplex) 형식의 단위세대를 적층해 4개 층의 전 세대가 지하와 옥상에 서비스면적을 갖도록 함으로써 효율적인 공간활용뿐만 아니라 서로가 마주칠 수 있는 가능성을 촉발시키도록 디자인되었다. "여지의 디자인"을 통해서 공공성을 다루는 것은 우리의 디자인 전략의 핵심이다.

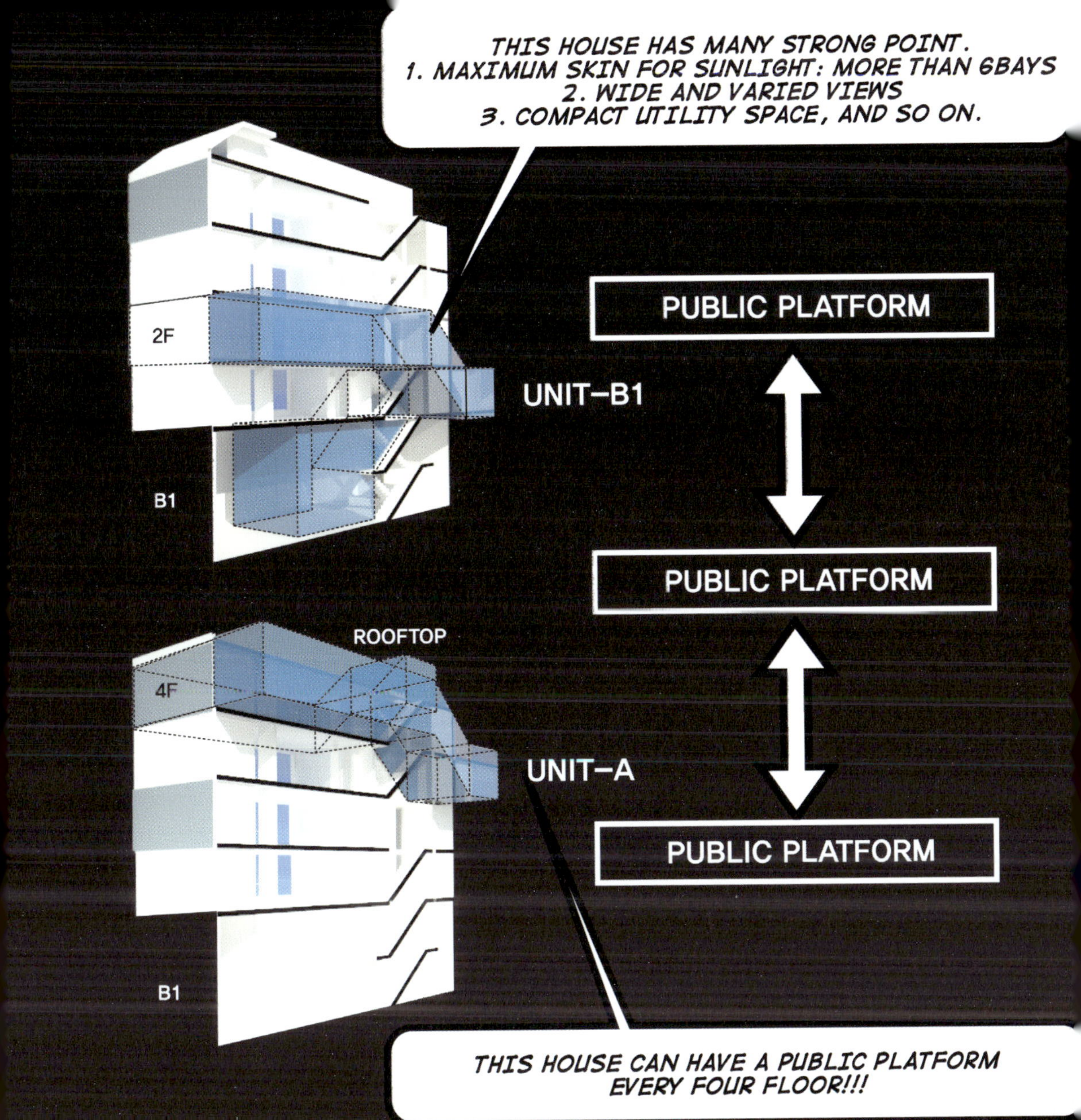

In brief, this project has been designed in proposal of individual units that were stacked four stories high in a skipped duplex style, where every household has a service area in both underground and the rooftop so as to activate not just efficient space-use but also possible encounters with each other.

It is our crucial design strategy to deal with publicness through "design of affordability".

HOUSE CROSSOVER IS A PROTO-TYPE OF SUPERPOSITION

SUPERPOSITION-TYPE IS DESIGNED TO MAKE MEDIATING HETEROTOPI
OF WHICH INDIVIDUAL UNITS ARE SEAMLESSLY CONNECTED
WITH PUBLIC SPACE SUCH AS A PARK OR A STREET

PUBLIC STREET
FLOATING PARK
CE,
PARK

DESIGN

ACTIVATING HORIZONTAL BOUNDARY

Extruded Community

문정동 공공원룸임대주택

Munjeong One-room Public Rental Housing

문정동 공공원룸주택은 길을 품고 도시로 열린 공동주택 프로젝트다. 장방형의 좁고 긴 대지의 특성을 고려하여 건축물 전체를 상부로 들어올려 주차와 보행로를 확보하였고, 커뮤니티 시설을 각각 북측 전면가로변 2층(세탁실)과 인접세대와 면한 남측 쌈지 공간 1층(커뮤니티센터)에 배치하여 보행로와 연계시킴으로써 대지의 안팎으로 깊숙히 이어지는 "공유공간"을 디자인하였다.

이와 같은 배치를 통해서, 표정없는 측벽으로 막히고 막다른 골목으로 갇힌 기존의 연립주택 배치와 차별화되는 열린 공간 구성이 가능해 졌으며, 남측 인접대지 다세대주택 주민들이 사용해오던 접근로와 작은 텃밭이 그 기능을 유지할 수 있게 되었다.

더 나아가, 세탁실이 위치한 2층의 외부 커뮤니티 휴게데크를 비롯한 다양한 폭과 깊이를 갖는 길과 복도, 그리고 옥상정원을 보행로와 면한 서측면에 수직적으로 중첩되도록 함으로써, "공유공간"이 입체적인 길을 품어 수직적으로 확장되고 활성화되도록 하였으며, 이를 통해서 커뮤니티 시설과 공간이 거주민은 물론이고 인근주민들에게도 쉽게 인지되고 접근되도록 의도하였다.

한편, 문정동 공공원룸주택은 극대화된 공간효율성 가운데 삶의 단면과 여지가 드러나는 공동주택 프로젝트다. 경제성과 사회성이라는 두 마리 토끼를 동시에 잡아야하는 임대주택의 특성을 고려하여 공간은 최대한의 효율성과 공공성을 목표로 계획되었으며, 효율적이고 유연한 공간활용을 위하여 확장과 변화가 용이한 라멘구조가 채택되었고 데드스페이스를 없애고 자투리공간을 수납공간 등으로 활용할 수 있는 다양한 공간장치가 적용되었다. 동시에 개성이 드러나는 집합적 형태를 모색하기 위해서 층별로 테마화된 공유 공간이 계획되었다.

1층의 커뮤니티 센터와 텃밭, 2층의 세탁실 및 휴게데크, 2,3층 복도의 투명한 수납벽(거주자의 라이프스타일을 표현하는), 4층의 공유 테라스, 5층의 다락과 옥상정원이 계획되었다. 이들 공간은 컴팩트한 단위세대 외부 경계에 여지의 공간을 형성함으로써 빽빽한 볼륨에 여유있고 매력적인 표정을 창출하고 있다.

Munjeong public rental housing is the project for embracing a street and opening to the city. Taking into account its narrow and long rectangular form, we lifted up the entire building to secure a parking and a pathway for pedestrian, and designed deeply penetrating "sharing space" which connects the inside and outside of the site, by allocating community facilities at both ends, along the pathway.
(public laundry room at 2nd floor by the street facing the north side, and community centre at 1st floor by the southern pocket garden facing surrounding buildings)

This layout allows more opened space organization to be built, differentiating from prevailing typical urban low house block characterized as being blocked by its expressionless sidewall and a dead end, and it helps maintain functions of an access route and small kitchen garden which local residents used to use in the southern adjacent site.

Furthermore, we were able to activate and extend the "sharing space" vertically, by superposing several types of pathways and corridors, such as a community deck facing a public laundry room on the 2nd floor, thereby making it embrace a three-dimensional street inside. Actullay, the supoposition was intended to make both residents and neighbors perceive and access the community spaces much more easily.

While, Munjeong public rental housing is the project for disclosing sections of life and room for life out of maximized spatial efficiency. Considering the economic feasibility and sociality required for a public rental housing, we approached the project to secure more effiecient space and the greater publicness as far as possible. For this, we have chosen Rahmen construction to increase spatial flexibility for extension or renovation, and developed spatial devices to remove dead space or to utilze leftover space for a storage or something else.
On the other hand, we have characterized an entire sharing space by a specific theme on each floor in an attempt to devise a distinctive collective form.

A community centre and a pocket garden on the 1st floor, a public laundry room and a community deck on the 2nd floor, a transparent storage wall on the 2nd and 3rd floor, which displays a resident's individual lifestyle, a sharing terrace on the 4th floor, and an attic and a roof terrace on the 5th floor are designed.
These spaces allow an attractive and relaxed expression on the tight volume, by making room for something else on the outer boundary of the compact house units.

건축가 말이
문정동 공공원룸주택은
함께모여사는 삶이 음각되고,
각자의 삶이 양각되어 빚어진
거대한 도시의 도장이래요!
어쨌거나 삶의 단면이 바깥에
잘 드러나기는 하네요^^
여기는 공용 세탁실^^
세탁을 돌리면서 책을
읽곤해요...
들쑥날쑥한 건물
모양이 재밌어요.
꽉 막혀 있지 않고,
다양하게 비워진
건물에 길도 열리고
숨도 트이네요.
항상 돌아서가던
친구네집으로 이제는
바로 갈 수 있어요.
집 뒤 텃밭에는
어제 엄마랑 상추도
심었어요!

IDEOGRAM

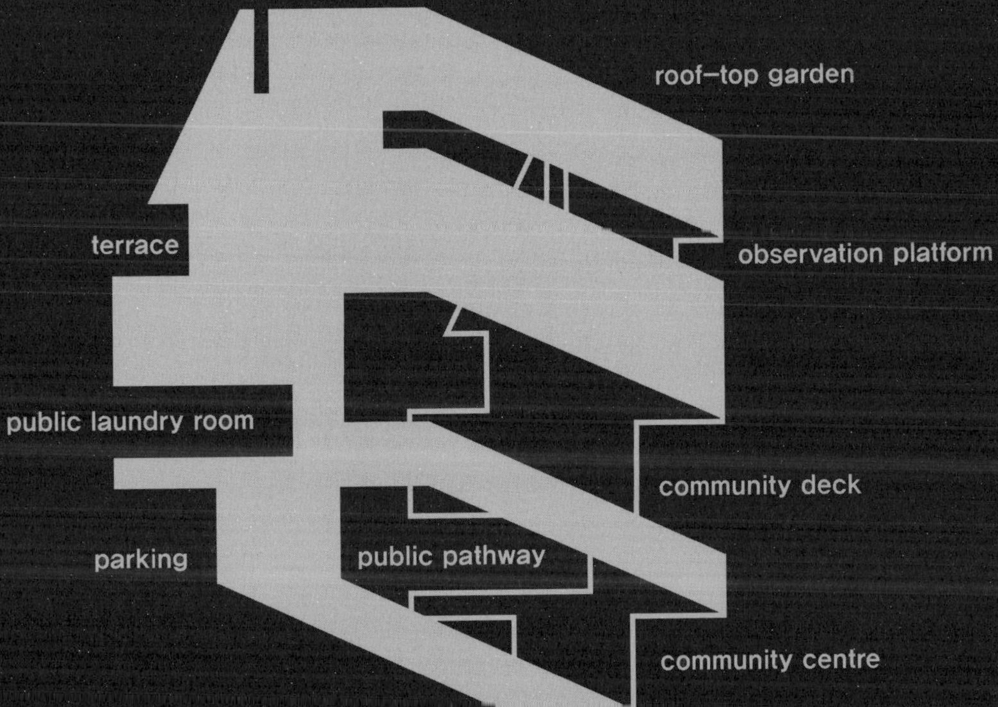
roof-top garden
terrace
observation platform
public laundry room
community deck
parking
public pathway
community centre

1 = 0 + 1

지속가능한 하나는 낭비의 제로 더하기 통합된 하나다.
A sustainable one is a waste zero plus an integrated one.

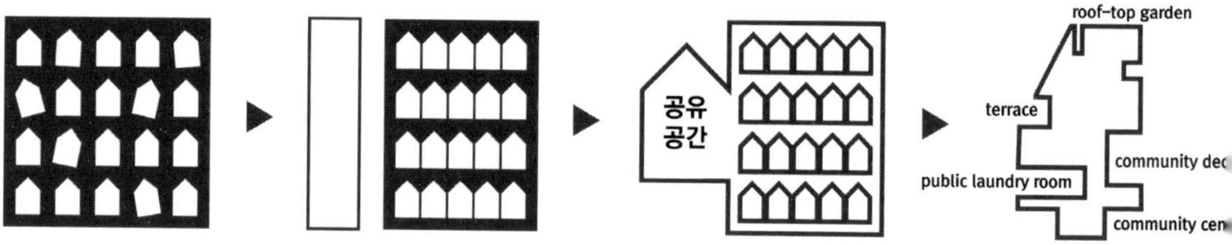

비효율적, 비사회적 공동주택

공간의 효율성 극대화를 통한 여유공간의 확보

여유공간의 공유공간화를 통한 지속가능한 통합설계

삶의 단편이 드러나는, 열린 공유공간 만들기

❶ 인근 주민들(대지 남측면에 인접한 다세대 주민)과 공유하는 공공보행로와 쌈지경작지 확보

❷ 건물 전체를 들어올려 지상층 전체를 커뮤니티 공간으로 개방 (공사비 절감 - 임대료 감소)

❸ 단지 내외부 경계에 거주민과 인근 주민들이 함께 사용하는 커뮤니티 거점 시설 마련

- 세탁실/서재: 가로변 2층
- 커뮤니티센터/AV룸/PC방: 대지 내부 1층

❹ 층별로 테마화된 공유공간 설계
- 1층: 커뮤니티센터
- 2층: 세탁실 및 휴게데크
- 3층: 락커룸
- 4층: 테라스
- 5층: 다락

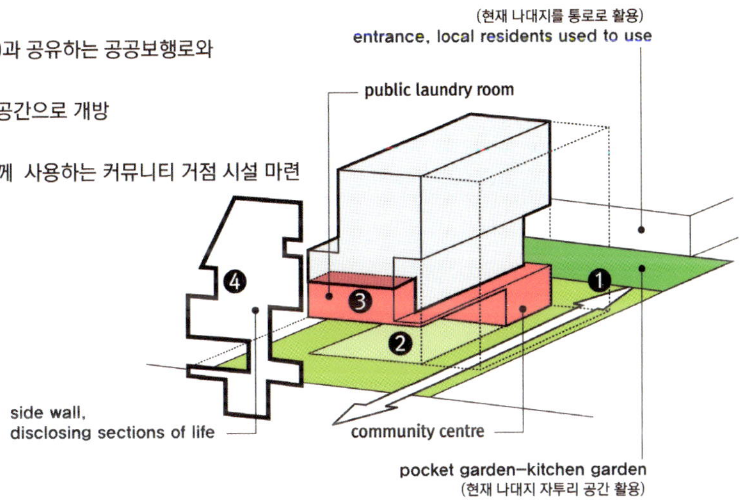

DRAWINGS

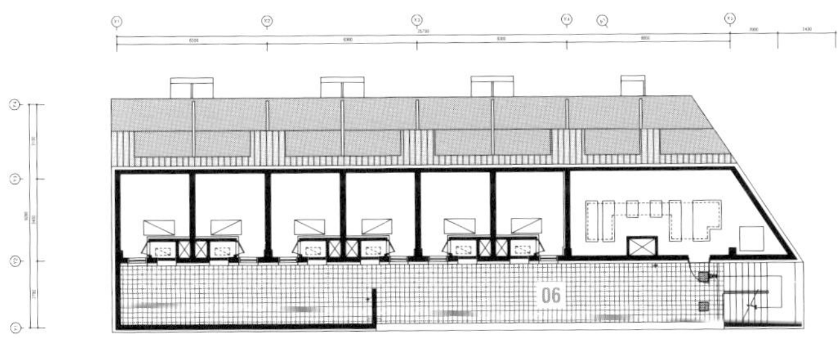

다락평면도

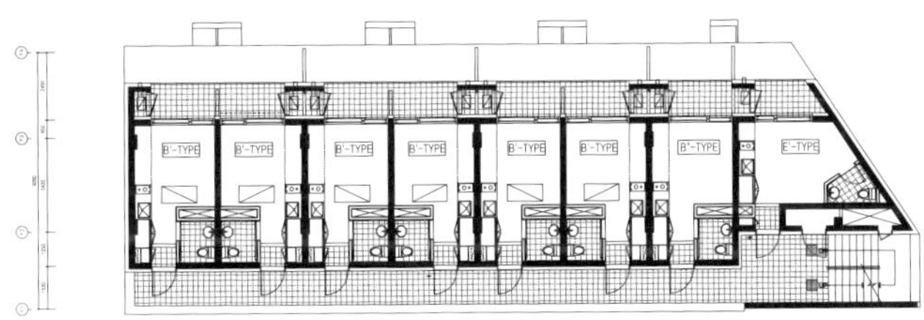

지상5층평면도

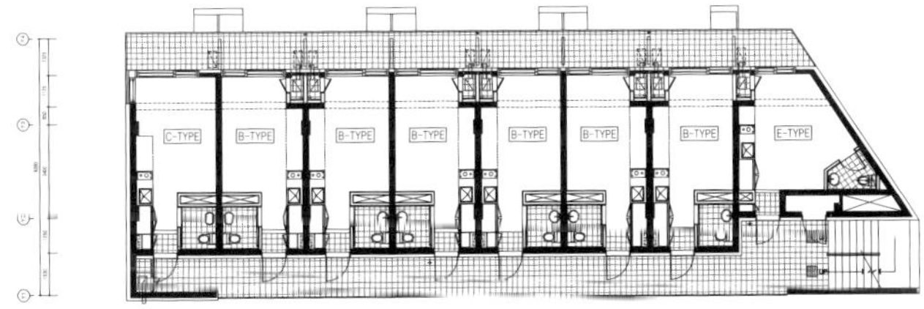

지상4층평면도

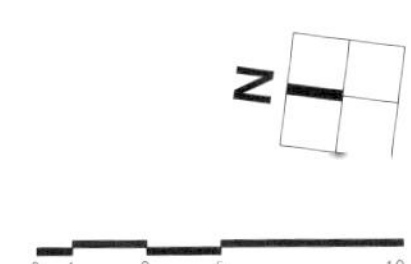

01_parking
02_community center
03_kitchen garden
04_community deck
05_public laundry room
06_roof-top terrace

DRAWINGS

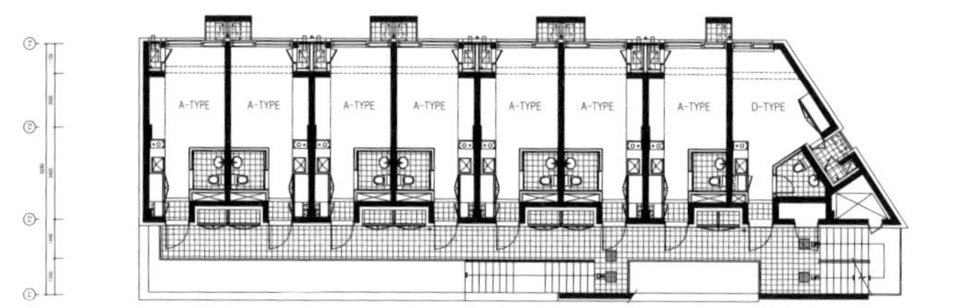

지상3층평면도

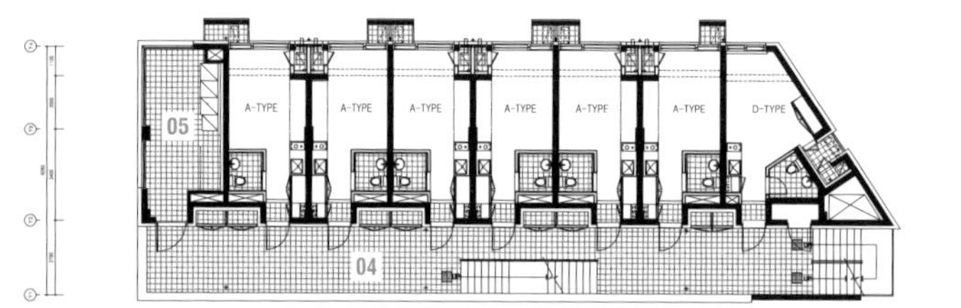

지상2층평면도

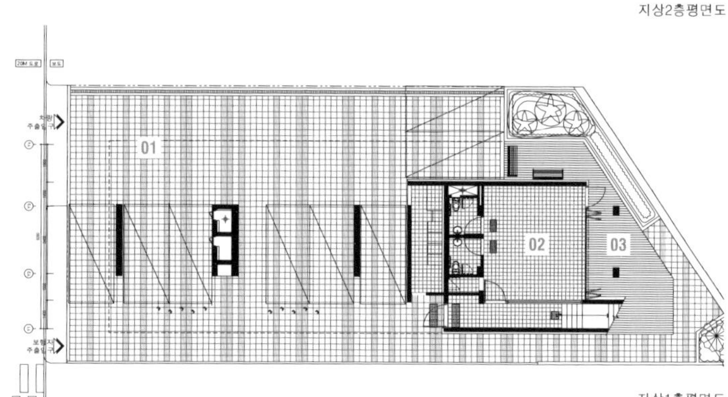

지상1층평면도

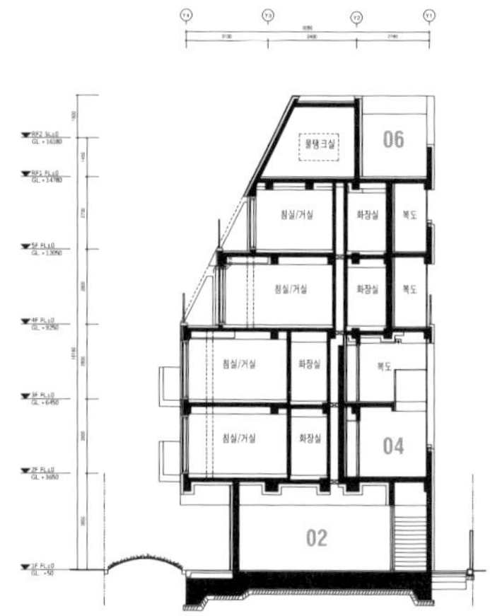

단 면 도 -1

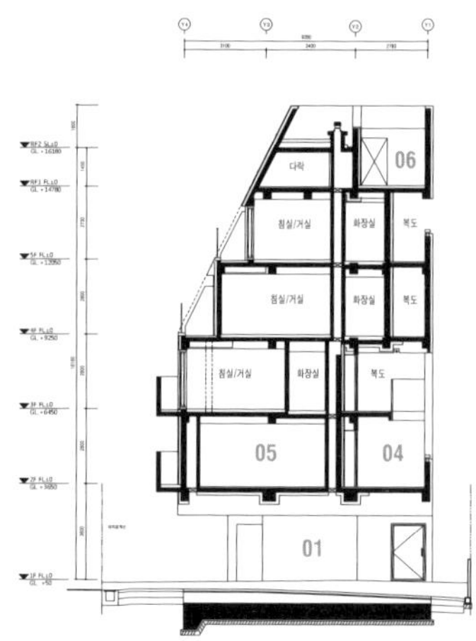

단 면 도 -2

문정동 공공원룸주택은
집합적 삶이 음각되고
개별적 삶이 양각되어 빚어진
압출성형된 커뮤니티,
랍니다!
담배꽁초버리지마세요.
IT SAYS
"PLEASE DON'T LITTER WITH YOUR CIGARETTE BUTTS"

"지속가능한 하나는 낭비의 제로 더하기 통합된 하나: 1 = 0+1"라는 개념하에, 우리는 공간 효율성을 극대화하는 컴팩트한 공간구성으로부터 여유공간을 확보한 후, 이 여지의 공간 혹은 비움의 공간을 만남과 소통, 휴식과 재충전의 삶을 담는 공유공간으로 변화시켰으며, 이를 외부에 표현하고자 하였다. 이로써 전체 건축물은 마치 음각된 도장과 같이, 하나의 솔리드한 볼륨에서 파내어진 다양한 형태의 공유공간이 드러나 삶의 단면이 곧 입면이 되는 공간이 되었다.

With the concept that "a sustainable one is a waste zero plus an integrated one: 1=0+1", we intended to make a freed up space out of an extremely compact and efficient spatial organization, convert this freed up or void space into a sharing space which embodies a life of encounters, communications, recreations and refreshments, and finally express it outside.
Thus, the entire building has turned into a space of which elevation is actually a section of life, disclosing various forms of a sharing space from a solid volume, just like a engraved stamp.

요컨대, 문정동 공공원룸주택은 집합적 삶이 음각되고 개별적 삶이 양각되어 빚어진 도시의 커다란 도장으로서 중층적 삶의 단면이 드러나는 압출성형된 커뮤니티(extruded community)다.

To sum up, Munjeong public rental housing is a big urban stamp where a collective life has been engraved and an individual life has been embossed, and an extruded community that expresses a section of multi-layered life.

SOLAR PHOTOVOLTAIC COLLECTOR PANELS
SIDE WALL DISCLOSES INTERATION BETWEEN INSIDE AND OUTSIDE
SHARING TERRACE
동시신호
PUBLIC LAUNDRY ROOM
LIFTED UP TO SECURE A PARKING AND A PATHWAY

HANGING CORRIDOR ABOVE A COMMUNITY DECK
TRANSPARENT STORAGE WALL DISPLAYS A RESIDENT'S INDIVIDUAL LIFESTYLE
CHEAP AND DURABLE FINISHING MATERIAL
PUBLIC PATHWAY WHICH LEADS TO NEIGHBORING RESIDENCES AND KITCHEN GARDEN

DESIGN

ACTIVATING HORIZONTAL BOUNDARY

House Cornerstone

동교동 근린생활시설+주택

Donggyo neighbourhood living facility+House

길모퉁이 건축

대지는 기존 주택가들이 상업시설로 변해가는 경계에 있다. 제각각의 개성과 관심이 충돌하는 주변의 상가건물들은 자유로운 분위기를 연출하기보다는 오히려 매우 혼란스러운 길거리 풍경을 만들고 있었다. 우리는 이에 맞서 건축물의 윤곽으로부터 단순하고 합리적인 경계를 찾고자 하였다. 이를 위하여 법규 안에서 최대로 찾은 연속적인 볼륨을 끌어내고, 이 볼륨을 다양한 공간들이 모퉁이를 돌아 서로 마주치고 연결되는 "맞물림 형태"로 변환시켰다.

Architecture at a street corner

The plot is located on the boundary where existing houses are being redevelopted into commercial buildings. These surroundung buildings makes a chaotic rather than a liberated streetscape with conflicting identities and interests of their own. Against it, we have searched for a simple and reasonable boundary out of a building's contour. For this, we have come up with a continuous maximum volume within regulations, and converted it into an "interlocking form", along which various spaces are interlocked and connected with each other around the corner.

이 건물 각층의 테라스들과 마당은 아늑하면서 곳곳의 다양한 풍경을 담고 있네요.
벽돌로 쌓은 외벽은 6,7층 주택을 둘러싼 마당의 담장이 되고, 상가 테라스의 난간이 되지요. 육중한 매스는 마치 공중에 떠있는 바위처럼 보이네요.
작고 친밀한 다양한 용도의 공간이 그 자체로 건물의 입면이 된 것 같아요. 삶의 단편이 음각된 듯한 형태가 도시 스케일의 아이콘이 되고 있는 것 같네요
테라스가 마치 코너를 돌면서 움직여 올라 가는 것 같아요. 두 길이 만나는 지점에 놓인 도시 표지판 같네요.

IDEOGRAM

URBAN TERRACE
AHEAD
!

DRAWINGS

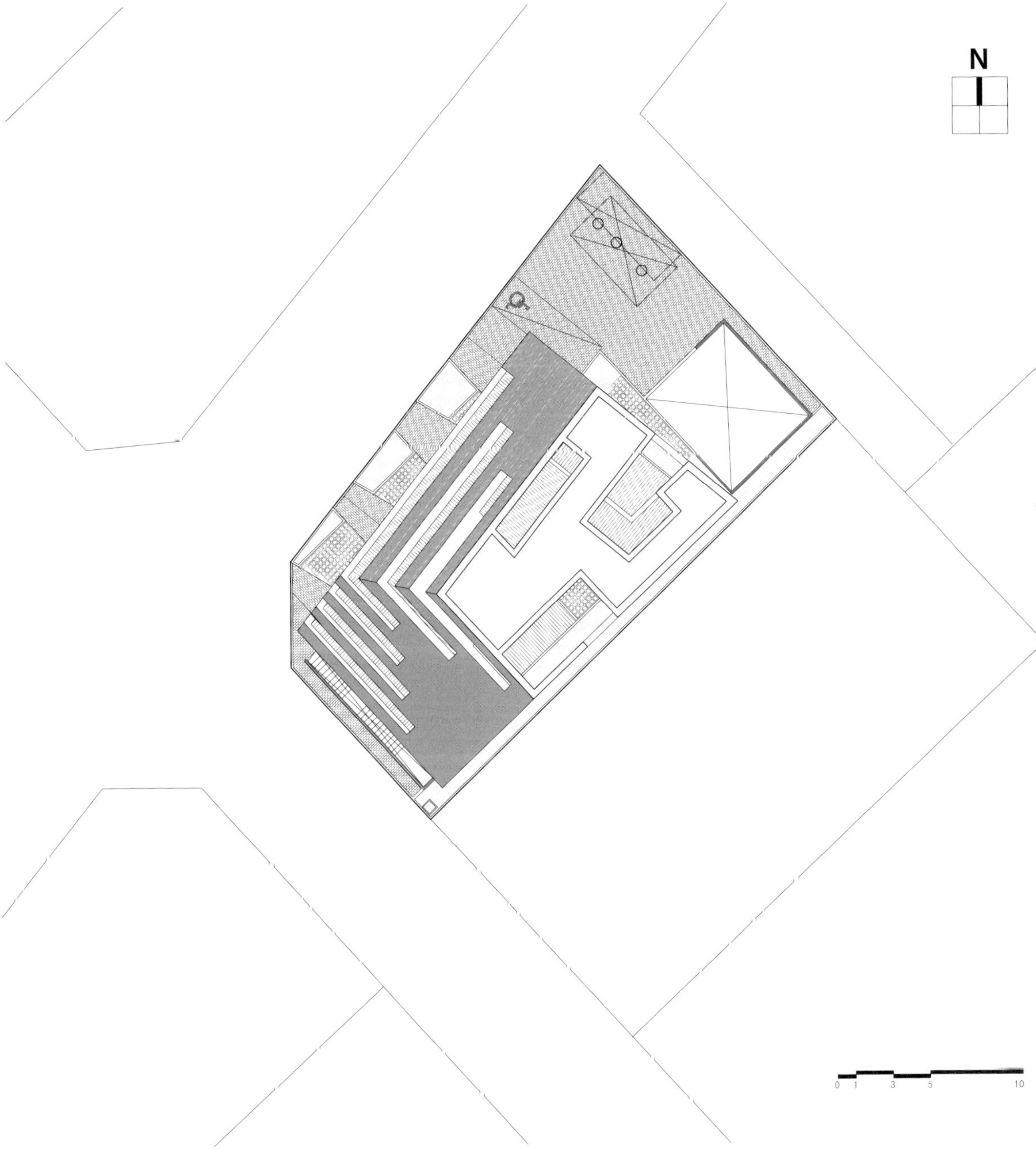

DRAWINGS

01_shop
02_deck
03_living room
04_master bed room
05_bed room

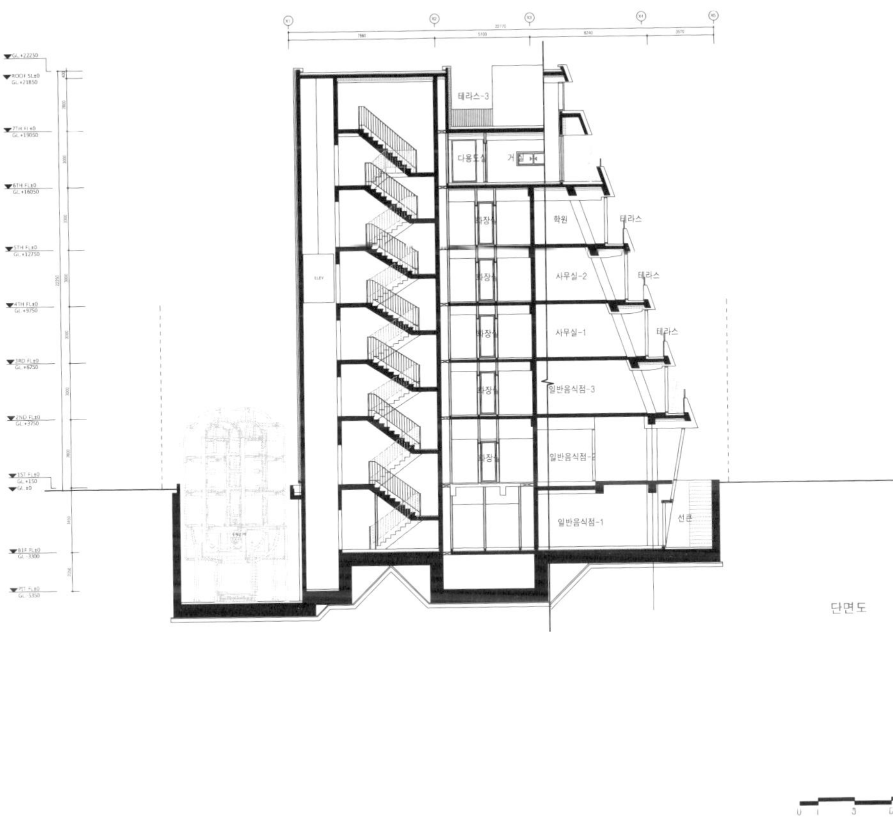

단면도

도시의 테라스

대지의 제약조건들에 의해서 유도된 볼륨의 사선면에는 층별로 특화된 도시적 보이드 공간을 제공하였다. 지상층에는 남측의 썬큰 마당과 도로변에 따라 생성된 야외공간들을 디자인하여 다양한 거리의 삶을 담는 여백을 제공하였고, 상부에는 경계면을 따라서 회전상승하는 테라스 공간을 계획하여 외부의 도시와 소통하고 내부의 삶을 드러내는 전이공간을 극적으로 표현하고자 하였다. 이로써 건물은 도시의 테라스로 둘러쌓인 극적 공간이 되었다.

Urban Terrace

Specific urban voids at each floor have been designed on the inclined side of the volume which has been driven by constraints and conditions of the site. On the ground, we allocated a sunken garden in the south and open spaces on the street, making room for various types of street lifes. While in the upper floors, we designed the dramatic urban terraces, an upspining transitional spaces along the boundary of the building, to communicate with the city of outside and disclose the life of inside. Thus, the building turns into a theatric space surrounded with urban terraces.

음각된 아이콘

6층부터 복층으로 계획한 옥상 위의 집에는 세 개의 작은 마당이 계획되었다. 이 마당들은 윤곽선 안쪽에 깊이있는 경계를 형성함으로써 프라이버시를 보장해주는 동시에 생활공간을 확장시키고 다양한 도시 풍경을 안으로 끌어오는 전이 공간을 제공하기위해 제안되었다. 우리는 이 마당들 역시 확장된 도시 테라스로 디자인하고, 마당을 포함한 모든 도시의 테라스들을 하나의 벽돌 외피로 쌓음으로써, 담과 벽의 구분이 없는 건축, 그리고 5개층의 상가와 2개층의 주택으로 구성된 전체 건물이 하나의 음각된 아이콘으로 보이는 공간을 만들고자 하였다.

동교동 상가+주택은 과다하게 장식된 중성적인 공간으로 전형화되어 있는 중소규모 복합개발에 대안적 방향을 제시한다. 이 프로젝트는 비교적 작고 친밀한 스케일의 일상적 생활공간을 극적인 방식으로 조직하여 개별적인 삶의 형상이 도시적 스케일의 아이콘으로 확장되는 길모퉁이의 건축이자 도시의 초석(CornerStone)이다.

Engraved icon

The duplex rooftop house at the 6th floor has three small gardens. These garden are planned to secure privacy by making deep boundary within the contour of the building, and to offer a transitional space that attracts various cityscapes inside and extends living space outside at the same time. Besides, we designed the gardens as the expanded urban terraces and covers all the urban terraces including the gardens, with a single brick skin, so as to come up with an architecture which does not have clear demarcation between a fence and a wall, and a space that makes a whole building, composed of 5-storey shops and 2-storey house, into a single engraved icon.

The Donggyo complex(shop+house) shows an alternative way for small and medium-sized Mixed-use development that has been typifed as an excessively decorated neutral space. This project is not just an Architecture at a street corner but also an Urban cornerstone which extends various individual life-forms to an urban-scale icon by organizing small and intimate everyday living spaces in a theatrical manner.

MAXIMUM VOLUME COVERED BY AN OUTER WALL,
EXPLORING THE BOUNDARY DRIVEN BY ARCHITECTURAL REGULA

ROOFTOP PRIVATE HOUSE + 3 PATIOS

IT EXPLORES THE OPPOSITIONAL INTERPLAY BETWEEN FORM AND FORMLESSNESS, SOLID AND VOID, ...

BRICK WALL

ENGRAVED VOIDS, LOOK LIKE ASCENDING

HERE COMES A STAIRCASE TO THE SUNKEN

분양
목화빌라
010-6342-4469
임대

ON

DESIGN

ACTIVATING DIAGONAL BOUNDARY

Street Pool Villa

남양주 주택-레스토랑-풀 콤플렉스

Namyangju House-Restaurant-Pool Complex

건축주는 북한강로변에 위치하는 대지에 레스토랑과 단독주택을 짓고 소규모 공연을 할 수 있는 공간을 갖기를 희망하였다. 분리된 기능과 필지 사이의 공간을 통합하여 묶어내고 북한강이 바라다 보이는 대지의 장점을 살려 계획하는 것이 하나의 커다란 숙제였고, 한편으로는 공적인 공간을 적절하게 배치하여 단지를 활성화 하고자 하였다. 먼저 단지의 대지 경계를 따라 커다란 담장으로 둘러 쌓아 전체 단지를 통합하고 전시와 이벤트가 가능한 공공갤러리를 두어서 공유공간으로 둘어쌓인 원형 극장과 같은 구성을 갖도록 하였다.

원형극장, 다양한 활동의 무대

대지의 특성상 합필이 어렵고 전체 연면적이 한정되었던 제약은 오히려 대지의 외부 공간을 풍부하게 해 주는 조건이 되었다. 레스토랑, 주택, 풀(pool) 등 사적 공간은 마치 공공의 바다 위에 떠 있는 섬들처럼 배치하고 공적인 공간은 대지의 자연스러운 경사를 이용하여 거대한 원형극장 공간(amphitheater)안에 다양한 형태의 쉼터와 무대 공간을 제공했다. 가장 낮은 곳에 무대와 같은 이벤트 광장이 위치하고 상부에는 사적인 풀이 위치하는 개인 정원이 그리고 상하부 공간을 자연스럽게 연결하는 두 건물 사이 공간은 객석과 같은 계단 정원이 위치한다. 이벤트 광장은 소규모 콘서트를 할 수 있는 무대공간이 위치하여 계단 정원, 레스토랑 하부의 계단 그리고 건물 내부 등 다양한 형태의 객석이 되어 관람할 수 있게 하였다. 또한 주무대 외에 내외부를 연결하는 담장의 개구부에 다양한 형태의 행위를 담을 수 있는 무대, 계단, 전망대 등을 설치하여 다양한 활동을 시각적으로 서로 공유할 수 있도록 하였다. 한편 북한강로를 두고 건너편에 위치하는 서룡개발 사옥은 건물 자체가 단지에서 바라보았을 때 북한강을 담는 하나의 프레임이 되도록 계획하였고 사옥 상부에서의 활동 역시 무대와 같이 연출될 수 있도록 했다.

The project owner hoped for a restaurant, a house, and a space for small-scale performance on the site located close to the road running along the Bukhangang(North Han-river).
The task was basically about consolidating seperated functions and the space in between the pieces of land, and planning to work with the strength of the site with the view of the river, while we also wanted to enliven the compound by appropriately arranging the public spaces.
More than anything else, a large fence is put on the perimeter of the site to unify the entire premises while a public gallery is placed to serve like an amphitheater surrounded by commons and sharing spaces.

The amphitheater as the stage for various activities

The restriction caused to the gross floor area by the characteristics of the site rather serves to enrich its exterior space. The private spaces including the restaurant, the house, and the pool are placed like islands floating on the site, while the public space takes advantage of the natural slope on the site to provide varied forms of resting places and a stage inside the large space of the amphitheater.
The lowest portion takes in the plaza for events including the stage, and the upper part supports a private pool harboring the private garden, while the space in between the building that naturally connect the higher and lower parts accomodates a leveled garden similar to audience seating.
The plaza for events holds the stage for small concerts, while the leveled garden, the steps in the lower portion of the restaurant, and indoor space supply diverse audience seating for an event. Also, at the opening that links the inside and outside, a stage, a staircase, and an observation deck are placed so that various activities can be visually shared with one another. Meanwhile, the Main office of Seoryung Development Co., located across the road, is planned to be part of one single frame that encapsulates the Bukhangang, while the activities on its top are intended to unfold like on a stage.

STREET POOL VILLA

IDEOGRAM

Pool
Restaurant
House
Event Plaza
Street Gallery
Street Theater

DIAGRAM

The amphitheater enclosed by rooms of activity
활동의 방(여지)으로 둘러쌓인 원형극장

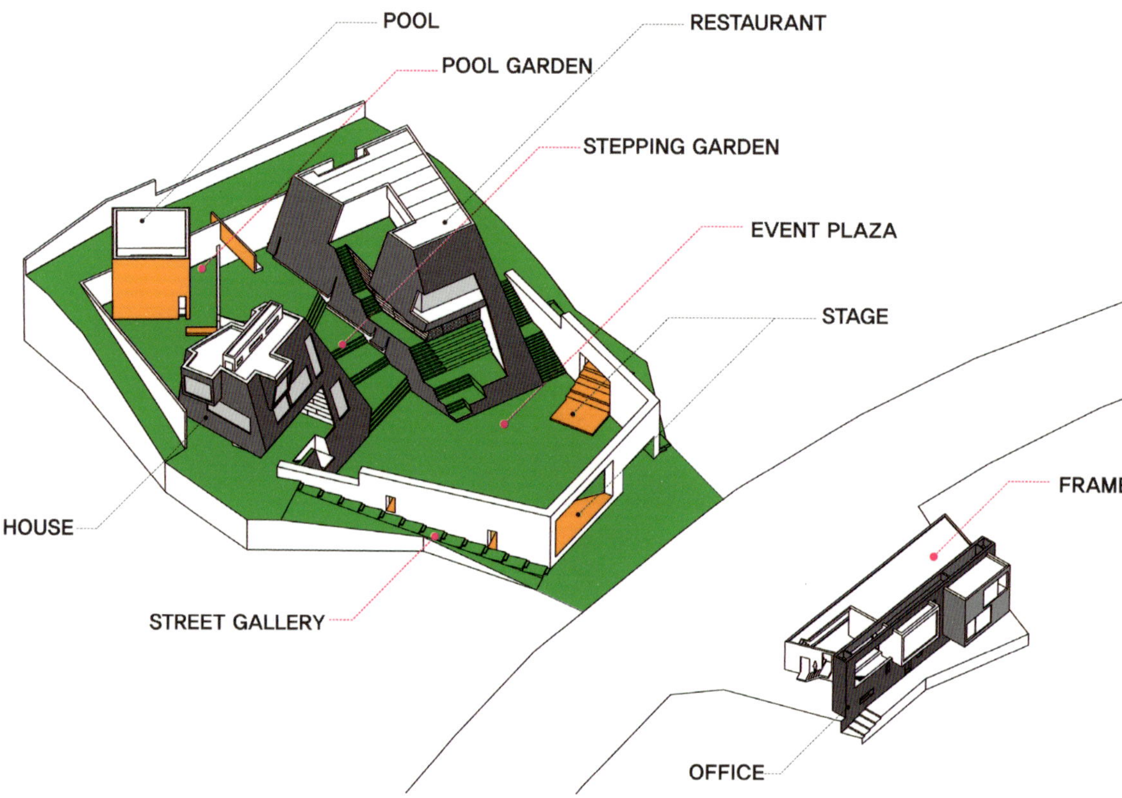

DRAWINGS

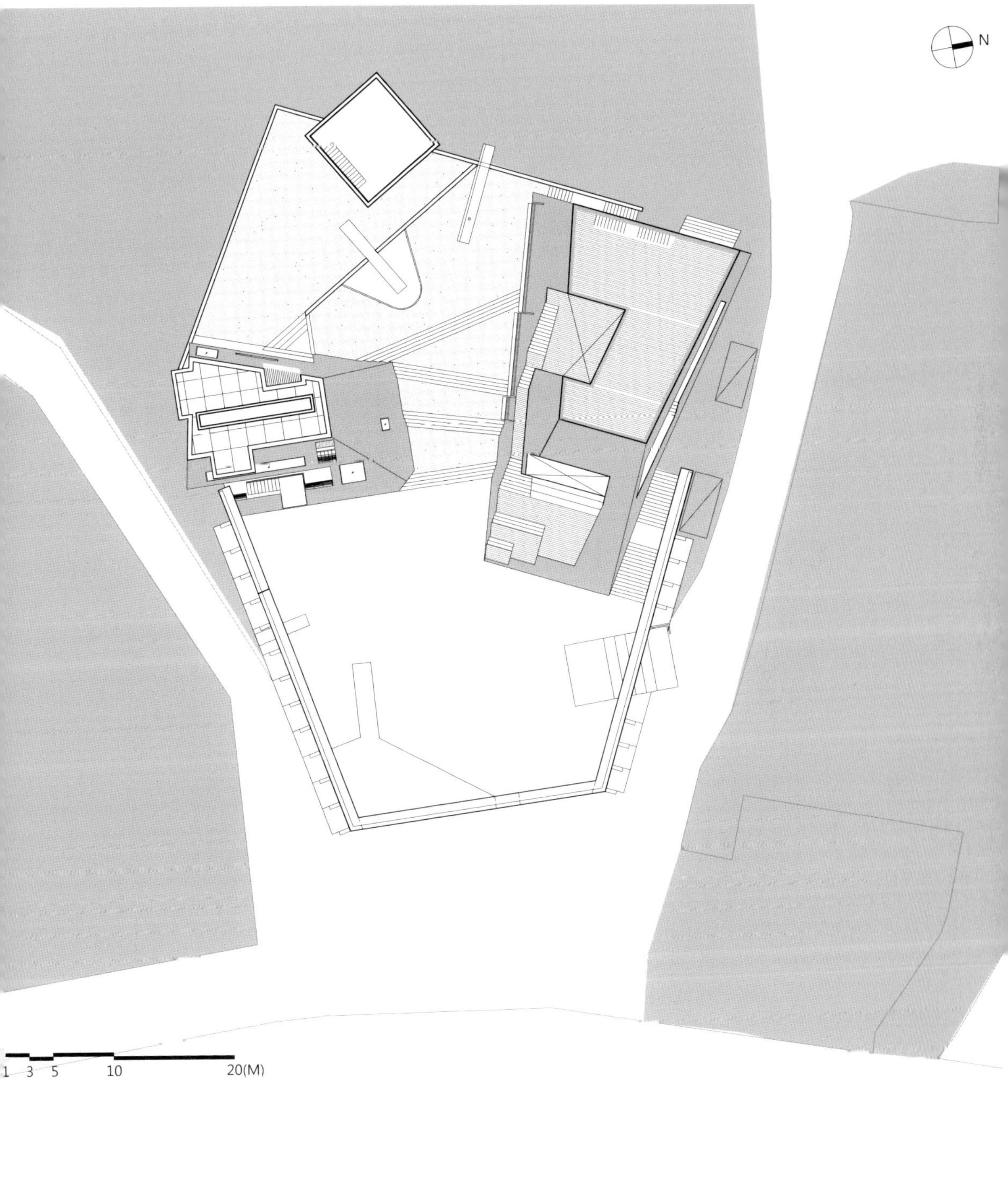

DRAWINGS
지상1층 평면도

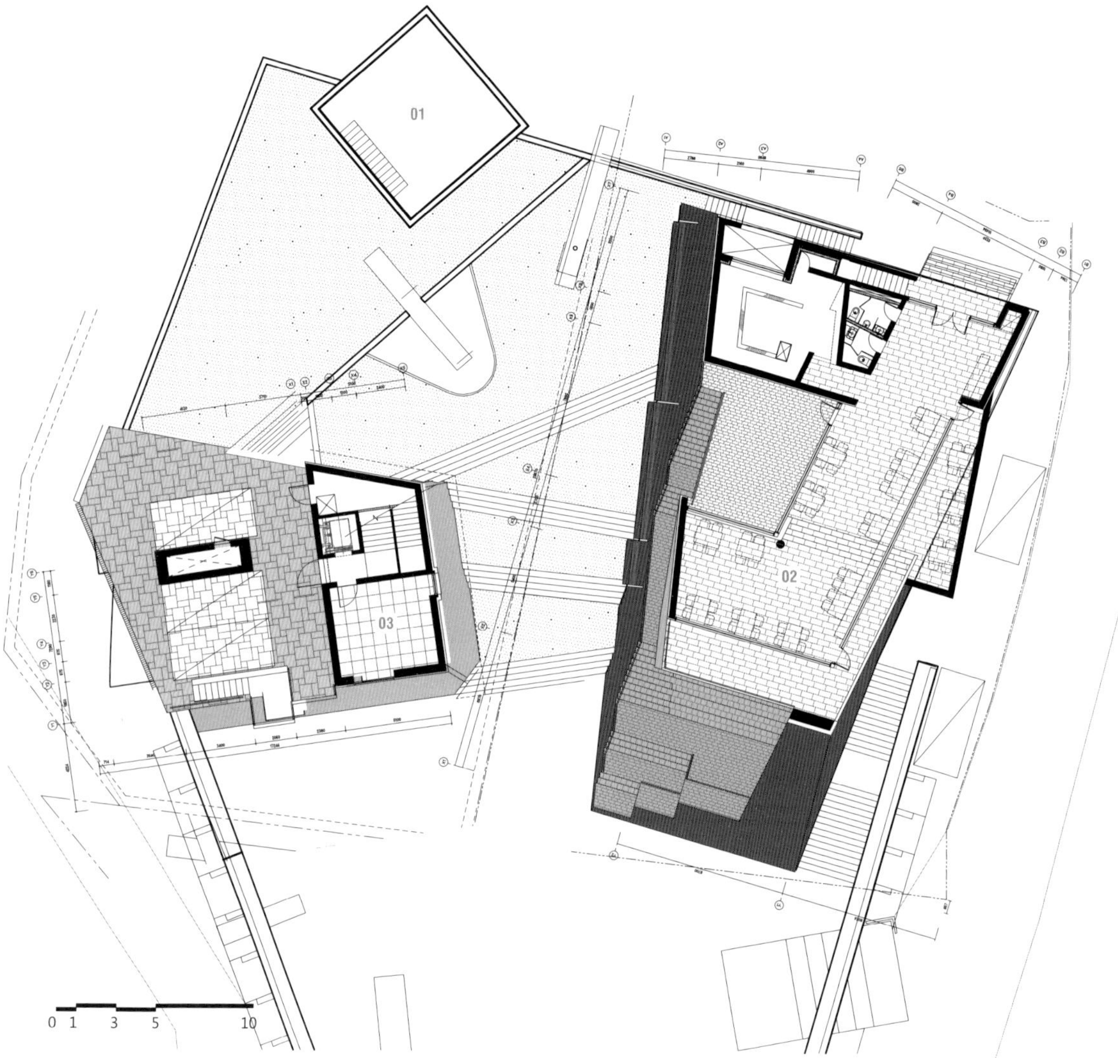

01_swimming pool
02_restaurant
03_event room
04_living room
05_master bed room
06_bed room

지상2층 평면도

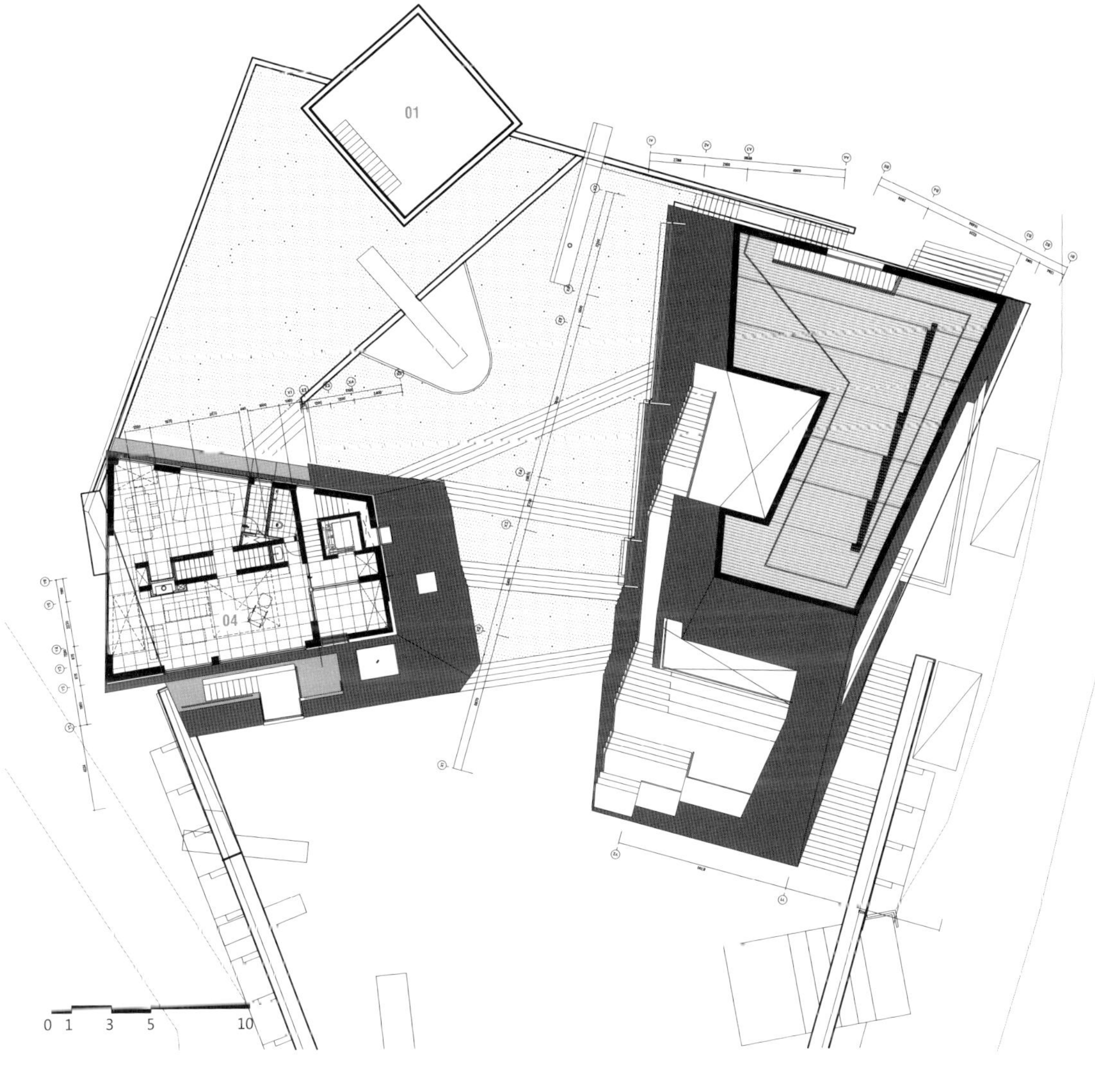

DRAWINGS
지상3층 평면도

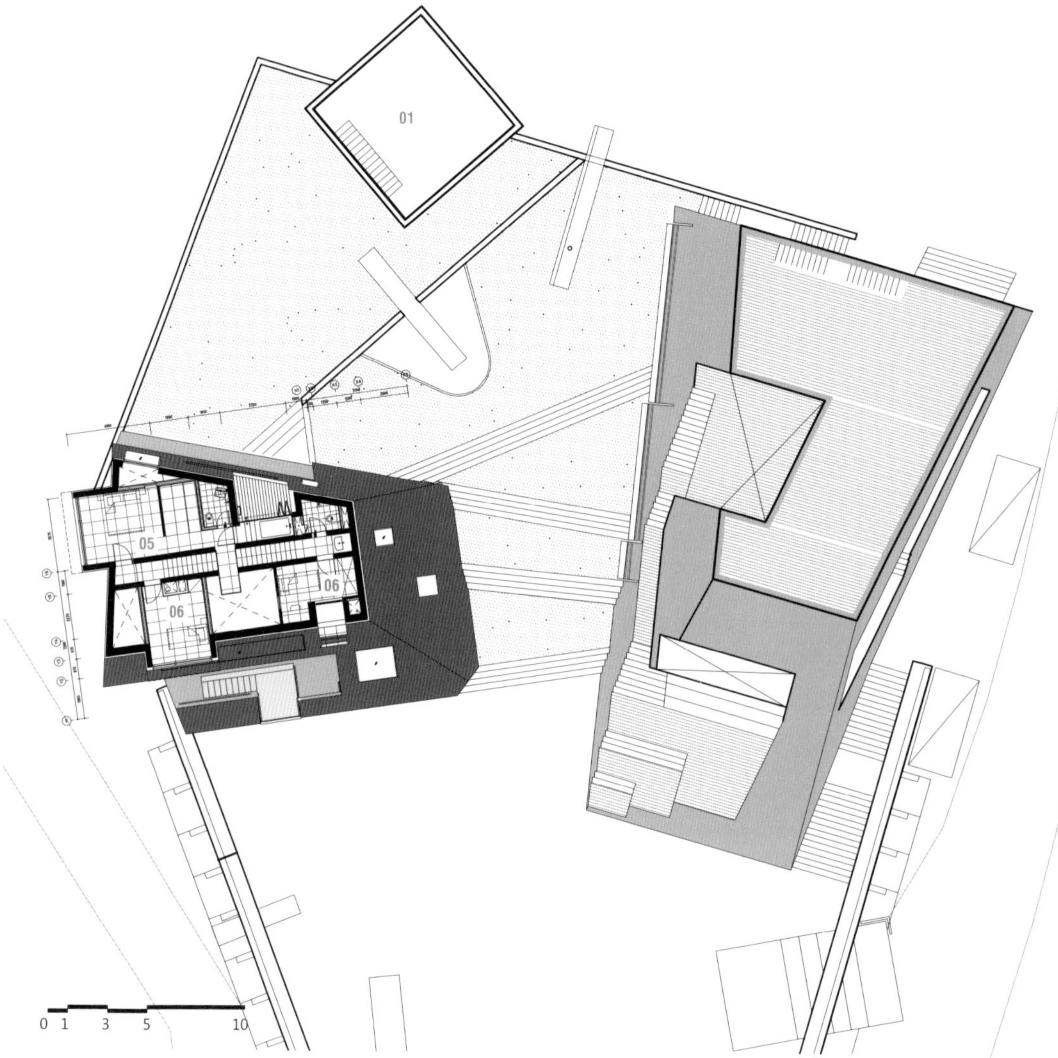

근린생활시설 단면도

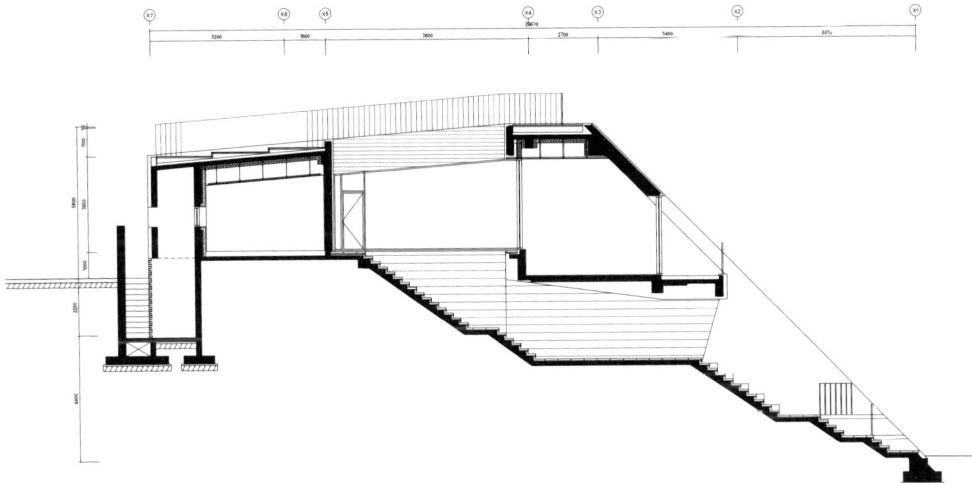

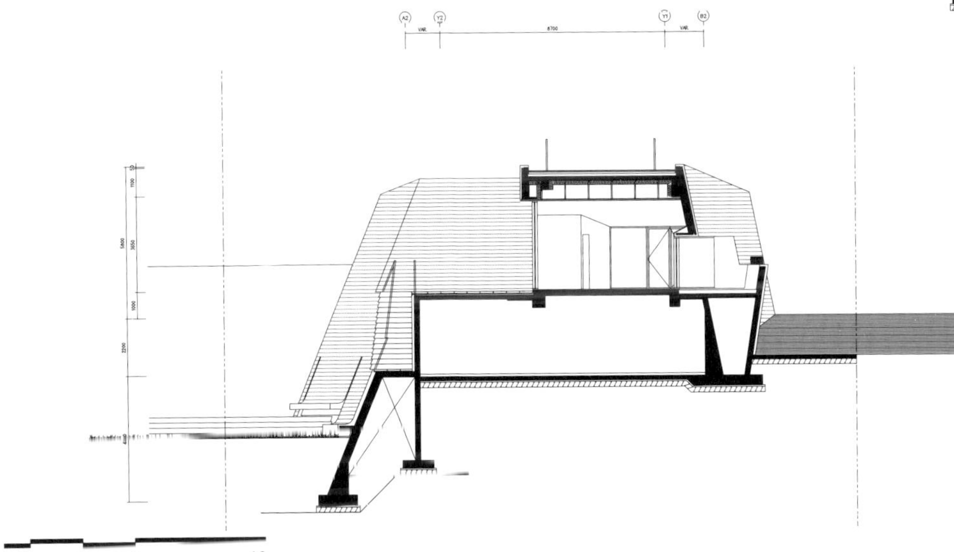

DRAWINGS
주택 단면도

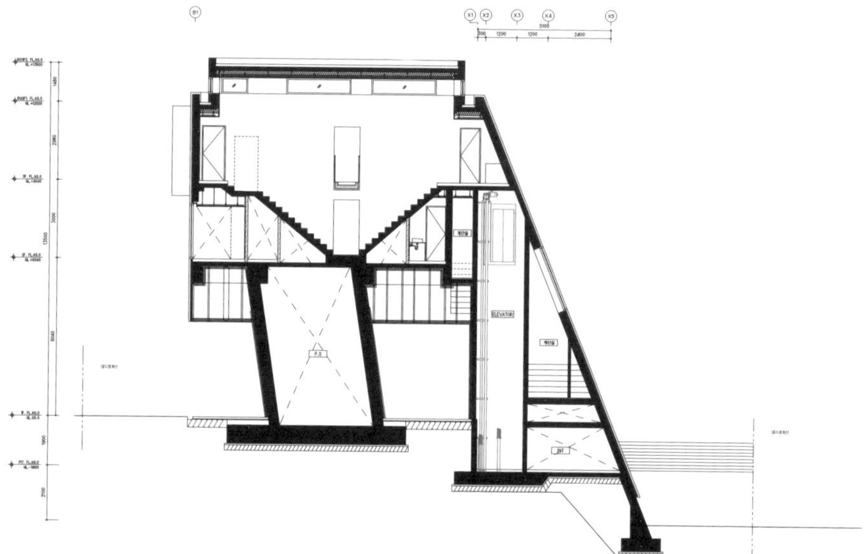

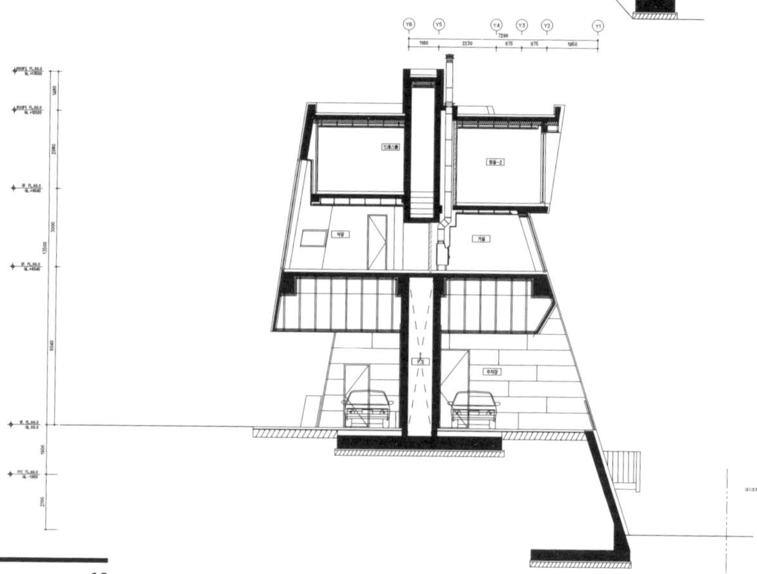

0 1 3 5 10

THE
H
TOWN
THE
H
TOWN

남양주 복합단지는 커다란 담장으로 둘러쌓여 있지만 폐쇄적인 공간이기 보다는 오히려 조망과 공간의 활용을 잘 조절한 공적인 공간으로 작동된다. 스테인레스 미러로 마감되어 주변의 풍경과 사람들의 행위를 비쳐주는 건물 하부면 또한 이러한 공적인 특성을 잘 보여준다.

The Namyangju Complex, while it is encircled with a large fence, does not come as a closed space, but it rather works as a public space with a nice coordination of the views and space use.
The lower part of the building that, finished with stainless mirrors, reflects the surrounding scenes and people, demonstrates its public character as such.

남양주 프로젝트는 담장 외부의 공공갤러리로부터 내외부를 관통하는 다양한 종류의 무대와 같은 공간들을 외부에 공유하여 공공성에 접근하고자 한 프로젝트이다.

The project in The Namyangju is designed to approach publicness by sharing with outside such various spaces like the stage that penetrates the interior and the exterior from the public gallery lying outside the fence.

PROJECTING ROOM
MIRROR WALL, REFLECTING
SURROUNDING ENVIRONMENT
MASSIVE AND DEEP BOUNDAR
CORDINATING OPENNING

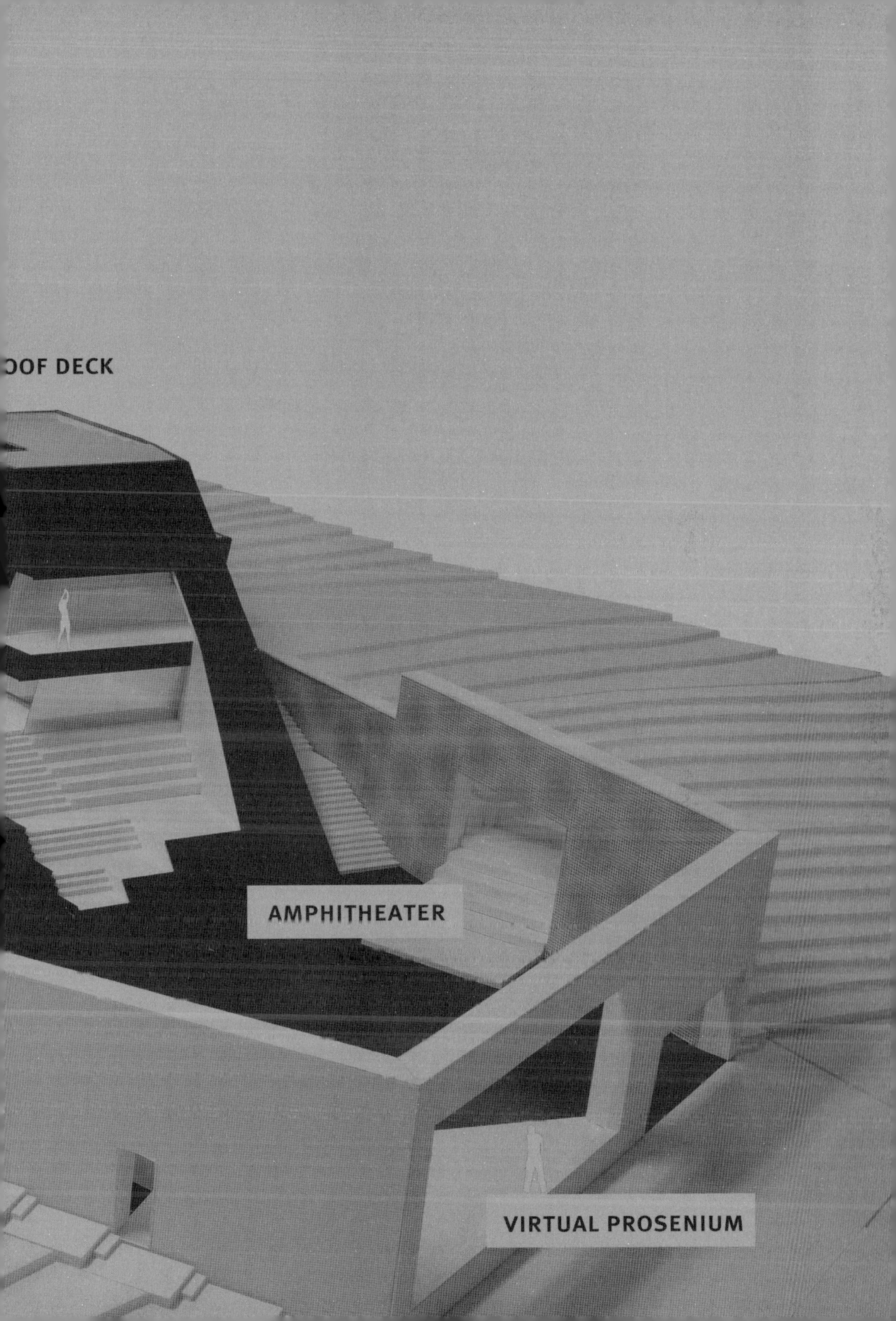
OOF DECK
AMPHITHEATER
VIRTUAL PROSENIUM

ACTIVATING DIAGONAL BOUNDARY

Street Housing

가로 타입 공동주택 계획안

Street-type Apartment Proposal

이 프로젝트는 2010년 베니스 비엔날레에 제출한 새로운 집합주거 프로토타입으로 설계되었다. 이 주택은 중첩, 병치, 그리고 맞물림-유형 가운데 병치 유형에 해당하는 것으로 공원이나 가로와 같은 공공공간과 각 세대의 반사적인 공간을 연결하는 연속된 전이공간을 만들어낸다.

This project has been designed as a new housing prototype submitted to the Venice Biennale in 2010. Among three different prototypes such as superposition-type, juxtaposition-type, and interlocking-type, it belongs to the "juxtaposition-type" that makes continuous transitional space which links every unit's semi-private space with a public space such as a park or a street.

STREET HOUSING

IDEOGRAM

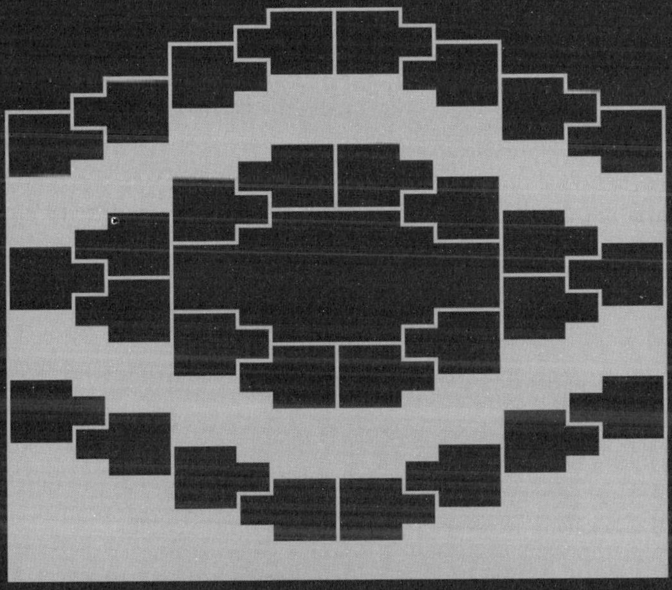

DIAGRAM

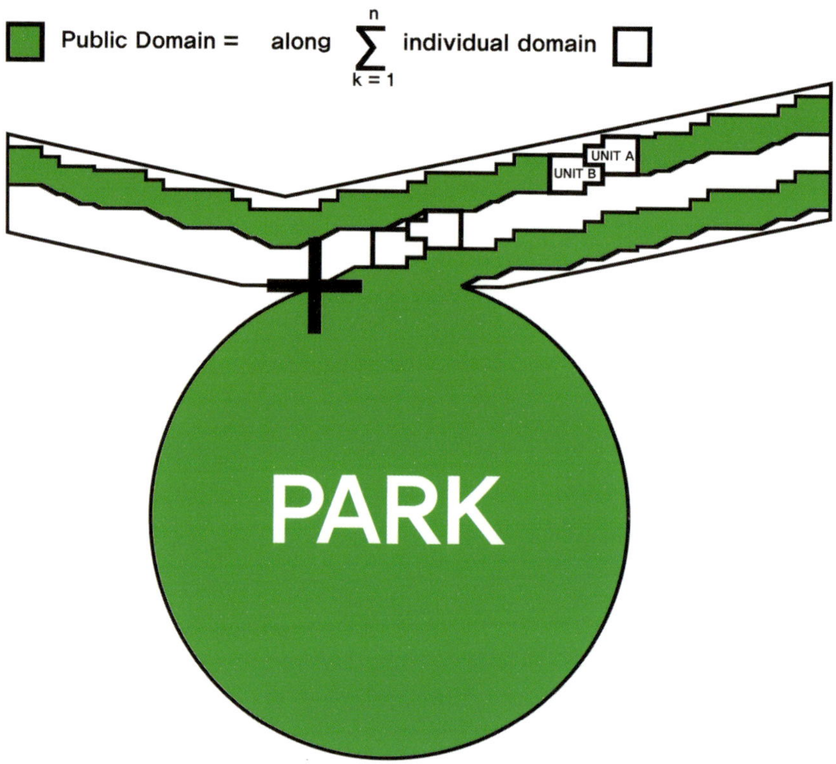

Public Domain = along $\sum_{k=1}^{n}$ individual domain
UNIT A
UNIT B
PARK

DRAWINGS

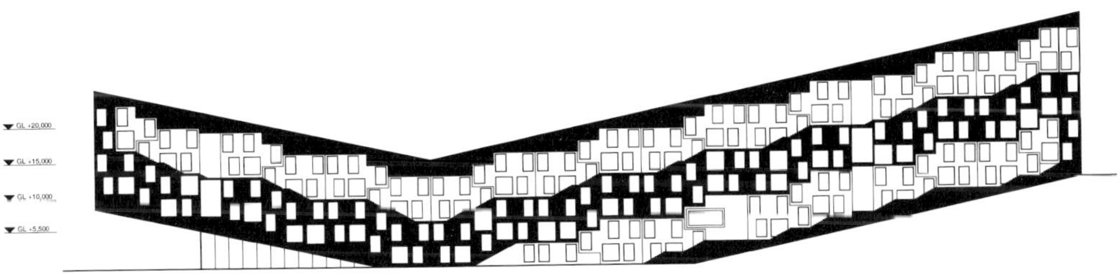

GL +20,000
GL +15,000
GL +10,000
GL +5,500

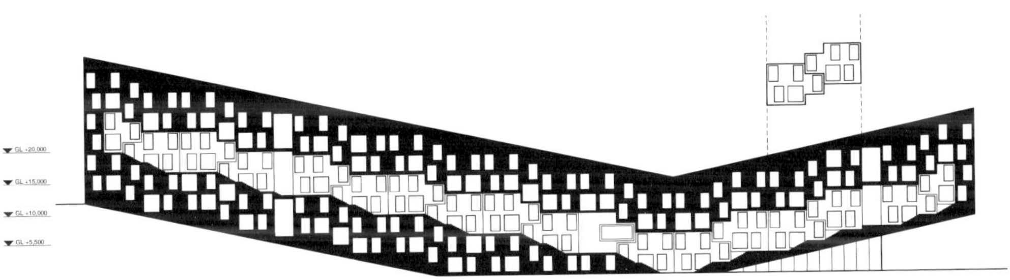

GL +20,000
GL +15,000
GL +10,000
GL +5,500

0 1 3 5 10

DRAWINGS

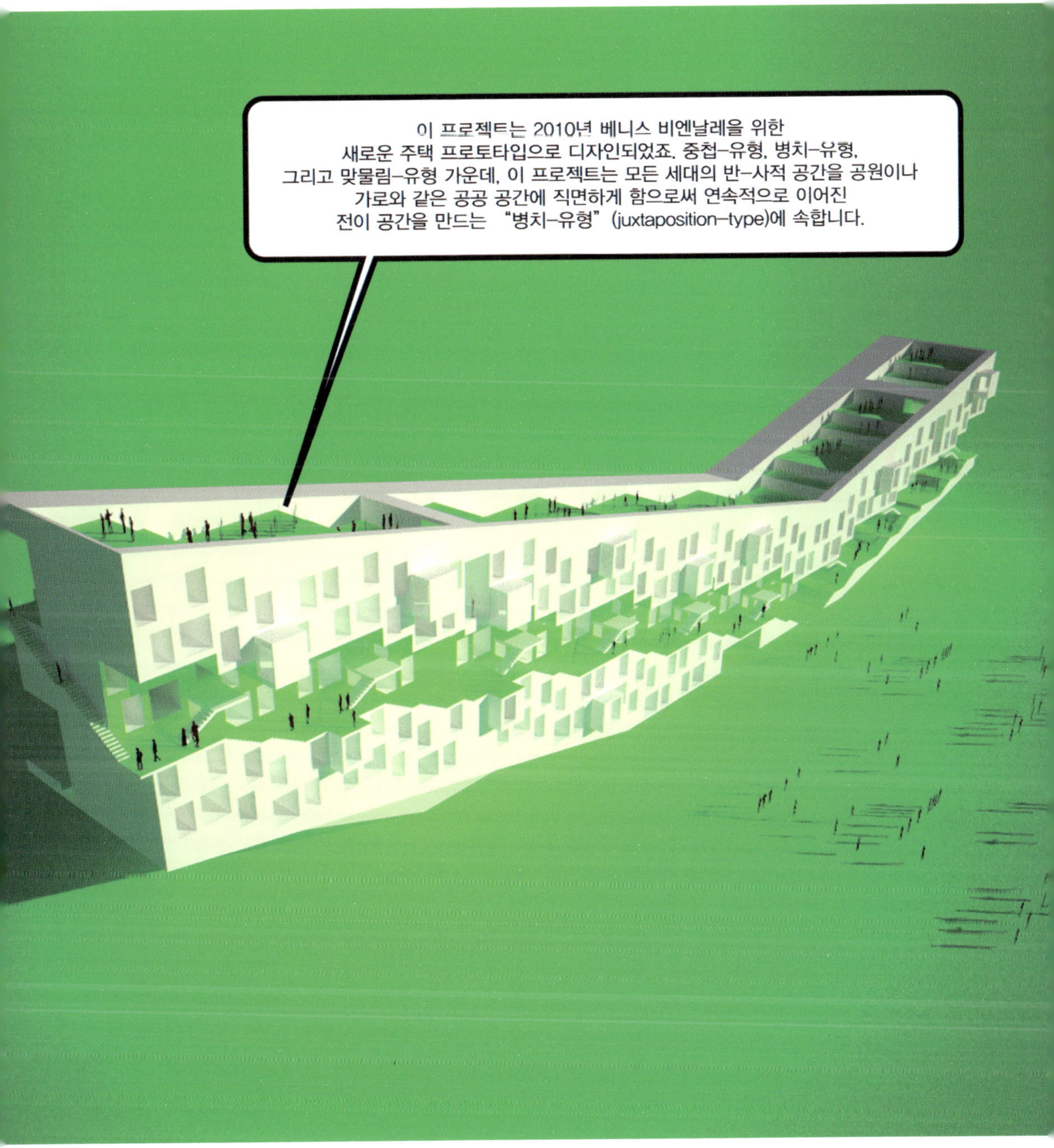
이 프로젝트는 2010년 베니스 비엔날레을 위한
새로운 주택 프로토타입으로 디자인되었죠. 중첩-유형, 병치-유형,
그리고 맞물림-유형 가운데, 이 프로젝트는 모든 세대의 반-사적 공간을 공원이나
가로와 같은 공공 공간에 직면하게 함으로써 연속적으로 이어진
전이 공간을 만드는 "병치-유형"(juxtaposition-type)에 속합니다.

이와 같은 레이아웃을 위해 듀플렉스와 스킵플로어의 장점 모두를 가질 수 있도록 결합하여 경사지에 쉽게 적용할 수 있는 하이브리드 타입의 주거 유형을 발전시켰다. 그리고, 각 주거유닛을 계단과 같은 형태로 차례로 병치하고 각 개별 발코니는 복도로 확장되어 관통하여 연속된 공적인 정원이 각 주호를 연결하도록 구축했다.

To make this layout, we developed a hybrid-type household unit, which can adapt itself easily to a sloping topography of the site, by combining a skip-floor with duplex one to take full advantage of the both. Then, those household units are placed side by side (juxtaposed) in a step-like form, and each individual balcony is extended to a hallway, such that an interpenetrating public garden was constructed to connect every household.

이 아파트 블록은 경사진 언덕 위에 기울어져 있는 것처럼 보이고 관통하는 공적인 정원은 계단 형태로 지면의 공원과 연결되어 발코니와 복도, 개별과 집합, 건축과 도시, 그리고 자연과 인공의 경계를 흐리고 있다. 그리고, 완충 공간이나 분리시설 없이 전이공간이 사적인 상태에서 공적인 상태로 또는 그 반대로 변환시키고 있다.

This apartment block looks like sliding on the slopping hill, and the interpenetrating public garden steps down to the ground park, blurring the boundary between the balcony and the hallway, the individual and the collective, the architectural and the urban, and finally, the natural and the artificial. Therfore, there is no buffers or seperators, but a transitional space that change from private state to public state and the other way around.

STREET_HOUSING IS A PROTO-TYPE OF JUXTAPOSITION
JUXTAPOSITION-TYPE IS DESIGNED TO MAKE MEDIATING HETEROTOPIA
OF WHICH INDIVIDUAL UNITS ARE SEAMLESSLY CONNECTED
WITH PUBLIC SPACE SUCH AS A PARK OR A STREET

CE,
PUBLIC STREET - PRIVATE GARDEN
COMMUNITY CENTER
PARK

DESIGN

ACTIVATING VIRTUAL BOUNDARY

Public Garden House

대구 산격동 단독주택

Daegu Sangyeok Detached House

건축주가 자택의 증축 설계를 의뢰했을 때, 유명건축가 승효상씨가 설계하여 거의 공사를 마친 티룸과 연결된 기존 주택을 리노베이션하고 최근 매입한 기존 대지 건너편 정원부지의 증축을 요구하였다. 따라서, 우리는 세 개의 다른 주택(증축부분, 거의 공사가 마친 티룸, 기존주택)과 세 개의 다른 장소(두 개의 다른 정원과 그사이에 위치한 도로)를 다룰 수 있는 통합적인 해결책을 제안해야 했다.

처음에 건축주는 두 개의 정원을 독특한 주택 콤플렉스로 물리적으로 결합하기를 원하였고 기존 주택을 리노베이션하고, 건너편 새로운 필지에 다른 증축건물을 세워, 도로를 건너갈수 있는 다리로 세 개의 장소와 세 채의 주택을 연결하기를 요구하였다.

그러나, 우리는 새 필지의 주택 및 다리의 증축 대신 도로를 따라서 브릿지 형태의 주택을 계획하고 (내부공간은 더 컴팩트한 반면) 더 열린 외부공간을 만들기 위해 그 정원을 길 반대편에 위치시켰다. 다시 말하면, 하부로 물리적 접근을 제공하고 상부로 시각적인 접근을 제공하기 위해 선형의 주택을 들어올려 계획하고, 건물로부터 자유로와진 반대편 필지(일본정원으로 계획)에는 잔디와 꽃, 그리고 장식적인 나무를 심어서, 대지에서는 직접적인 공공의 접근이 가능케 하되, 바다위에 떠 있는 다리를 떠올리는 주택의 들어올려진 데크에서는 간접적으로 사적인 접근을 가능하도록 하였다.

이 제안은 처음에 다소 낯설어 보였지만, 건축주와 우리는 새롭게 계획되는 주택이 개인주택이 가지는 배타적인 친밀감과 환경보존과 사회적 책임 사이에서 균형을 갖추어야 한다는 요구를 만족하기 위해서 새로운 대안이 더 합리적이라는데 동의하였다.

When the client commissioned us to design a house extension, he wanted to renovate his old house which is connected to the nearly-finished extension(tea-room), designed by renowned Korean architect Seung Hyosang, and build another extension within the recently-bought plot of land which is located on the other side of the house.
Therefore, we had to come up with an integrated solution to deal with three different houses (our extension, nearly-finished tea-room, old house) and three different places(two different sites and a road which is located in-between).

At first, he wanted to physically combine two different plots into a unique house complex, thus he asked us to renovate the old house, build up another extension on a new plot, and connect three different places and houses with a bridge that should be passing over the road.

However, we thought that instead of building a second extension and a bridge, it would be much better to have a bridge-shape house along the road and to locate its garden across the road in order to make the outside more open, while the inside more compact.
That is to say, we wanted to design a linear house which is lifted up to provide physical access below and visual access above, and make the opposite site free from buildings but furnished with lawn, flowers and ornamental trees, as a Japanese garden, such that it allows direct public access on the ground whereas indirect private access on the lifted deck of the house, which reminds us of a floating bridge over the sea.

This option was a little bit strange though, before long did the client and we come to an agreement that it would be more reasonalbe alternative as it came up to his requirement that a new house should balance exclusive intimacy with environmental care and social responsibility.

이 단독주택은 공공의 정원이지만, 주택에 딸린 마당같은 공간을 품고 있습니다. 그래서 건물 자체가 도시의 경계와 맞물리는 전이공간의 역할을 합니다.
할아버지댁 정원은 신기해요! 집과 길 너머에 있는데 집에 바로 붙어있는거 같아요!!
이렇게 정원을 바라보고 있으니 마음이 편안해 지네요...
여기는 공공에게 개방되는 일본 정원이예요.
여긴 도로^^

IDEOGRAM

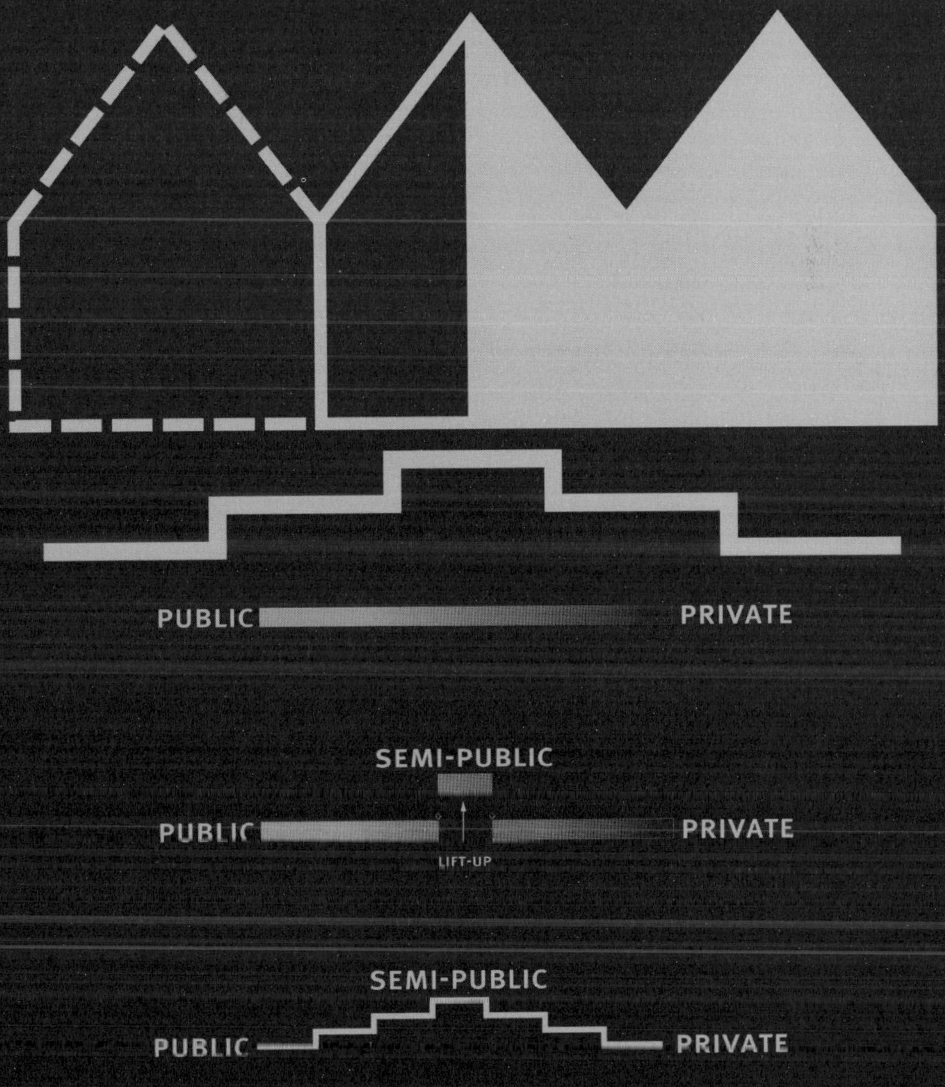
PUBLIC
PRIVATE
SEMI-PUBLIC
PUBLIC
PRIVATE
LIFT-UP
SEMI-PUBLIC
PUBLIC
PRIVATE

DIAGRAM

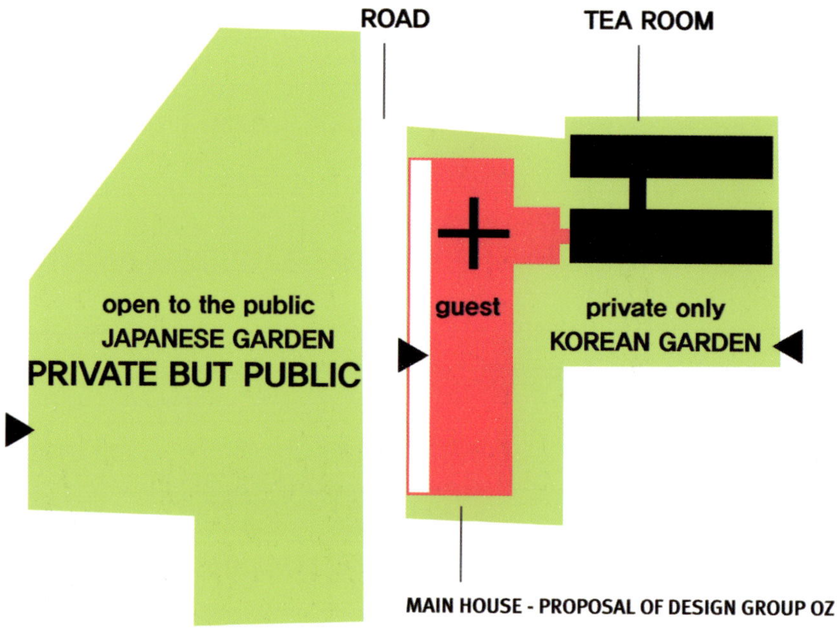
ROAD
TEA ROOM
open to the public
JAPANESE GARDEN
PRIVATE BUT PUBLIC
guest
private only
KOREAN GARDEN
MAIN HOUSE - PROPOSAL OF DESIGN GROUP OZ

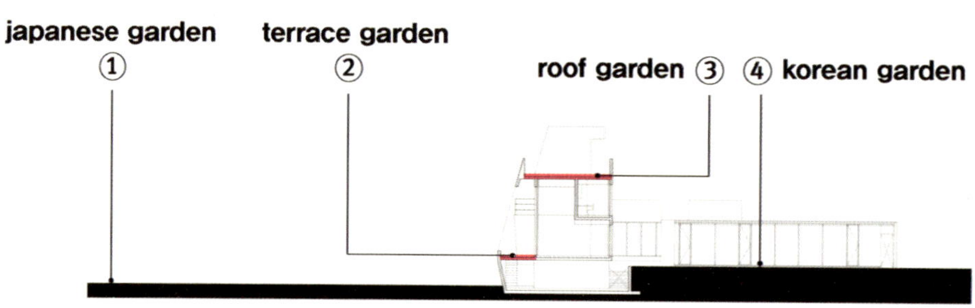
japanese garden ①
terrace garden ②
roof garden ③
④ korean garden

DRAWINGS

DRAWINGS

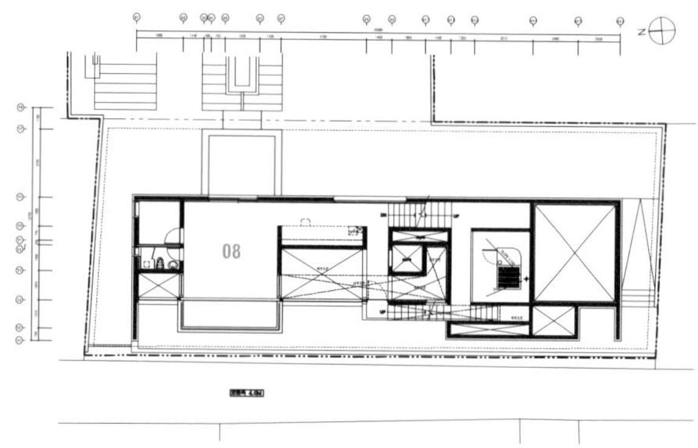

2층 평면도

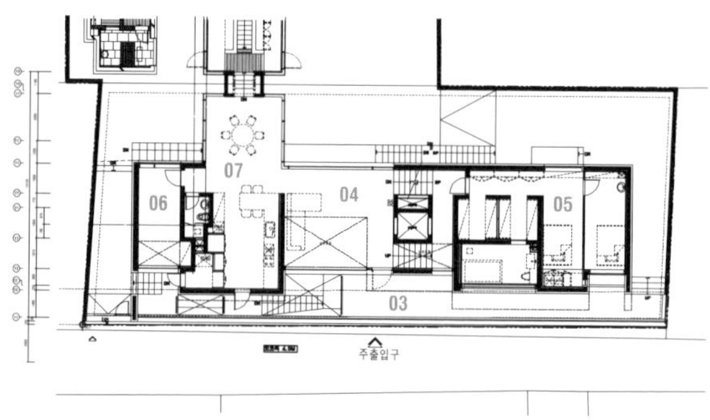

1층 평면도

지하1층 평면도

01_entrance
02_parking
03_deck
04_living room
05_master bed room
06_guest room
07_dining room
08_study room

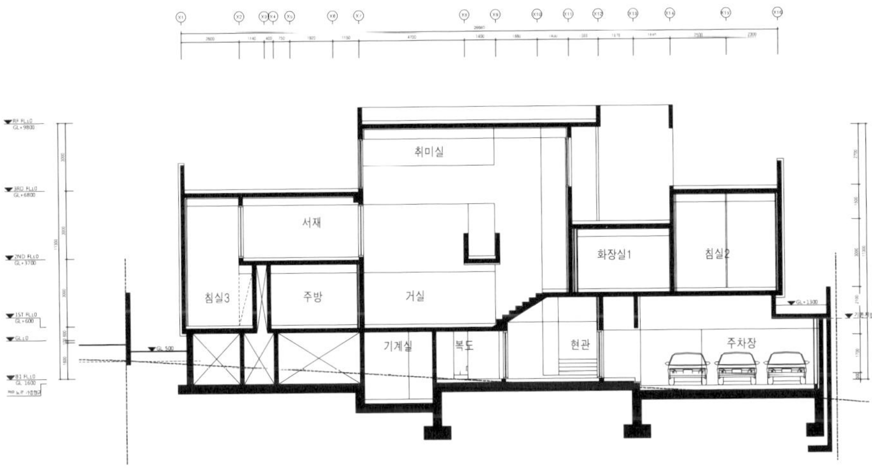

단면도-1

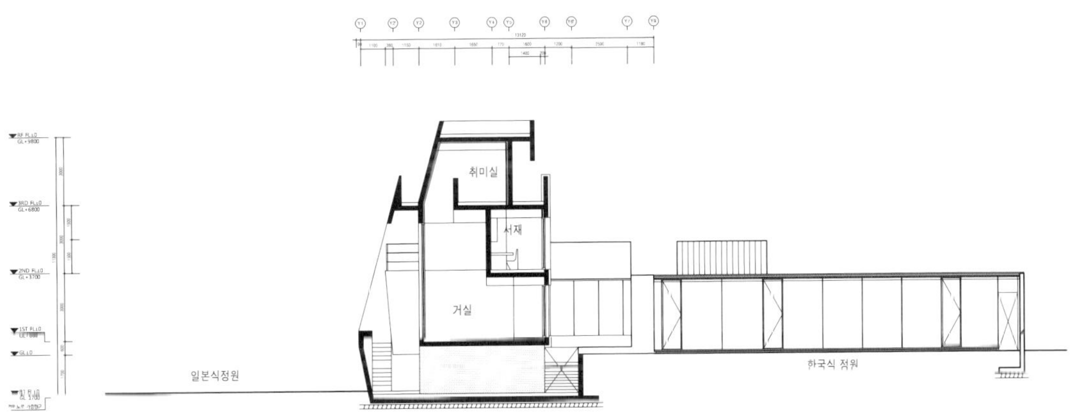

단면도-2

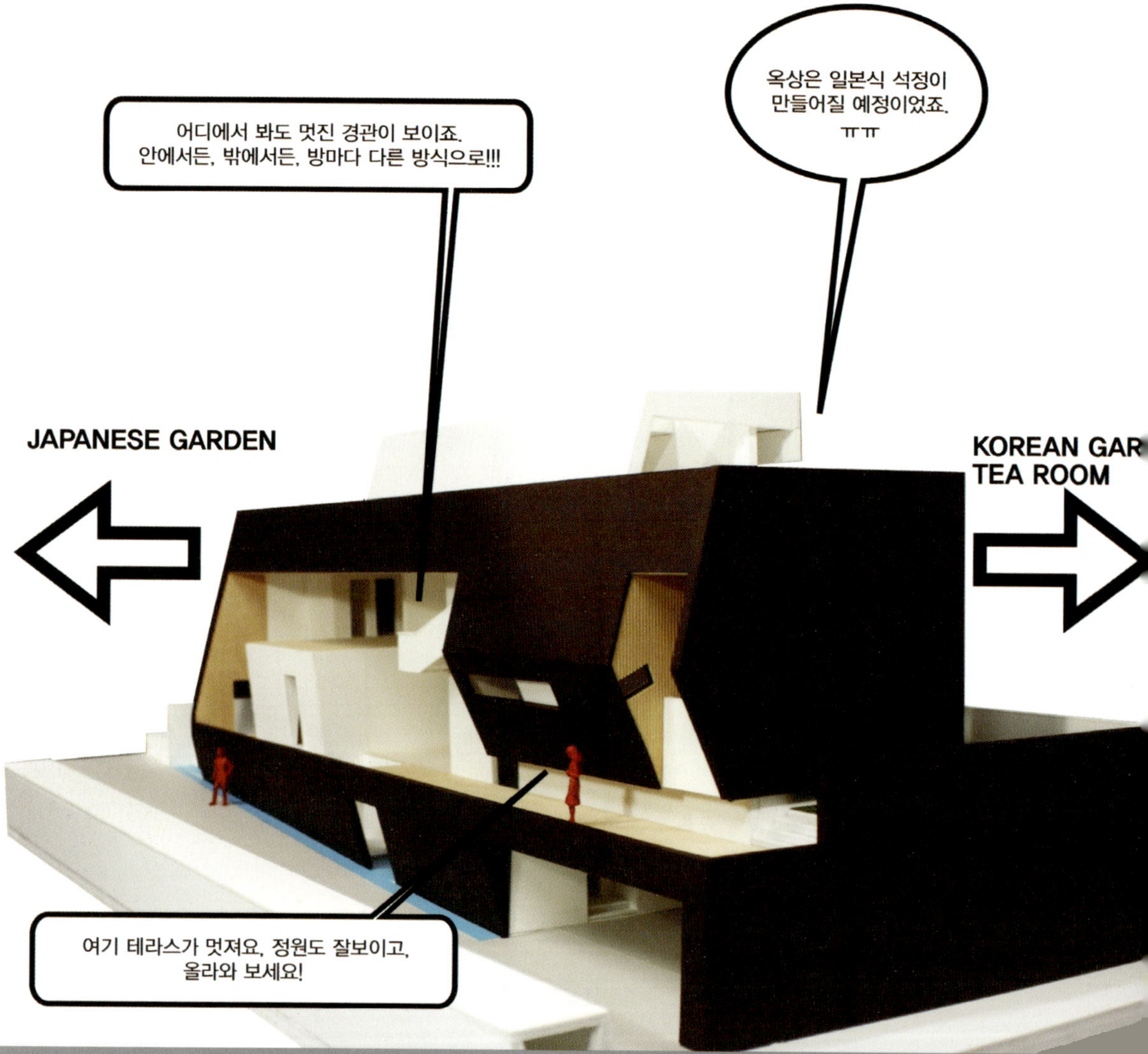

이 목적을 가지고, 우리는 입구 홀, 주차구획, 창고 등을 제공하는 플랫폼 위에 일본정원에 대면해 다양한 경관을 가지는 일련의 연결된 방들을 계획하고 그 중간에 식당 유닛과 티룸과의 교차점을 만들었다. 따라서 개인의 생활은 지면레벨에서 통행자의 시선으로부터 보호되는 한편, 일본정원을 둘러싼 담벼락 너머로 데크 위에 있는 각 방의 창들 앞에서 파노라마 경관이 펼쳐진다.

With this aim in our mind, we designed a linked series of room-units, which have different views of the facing Japanses garden and their own specific ways of viewing, on top of the platform that houses an entrance hall, a parking lot, utility rooms and etc., then, made an intersection to the tea-room wing with a dining room-unit, in the middle. Accordingly, the personal life would be kept from the sight of passers-by on the ground level, while, panoramic scenery would be unfolded before each room's window on the deck level over the wall that surrounds the Japanese garden.

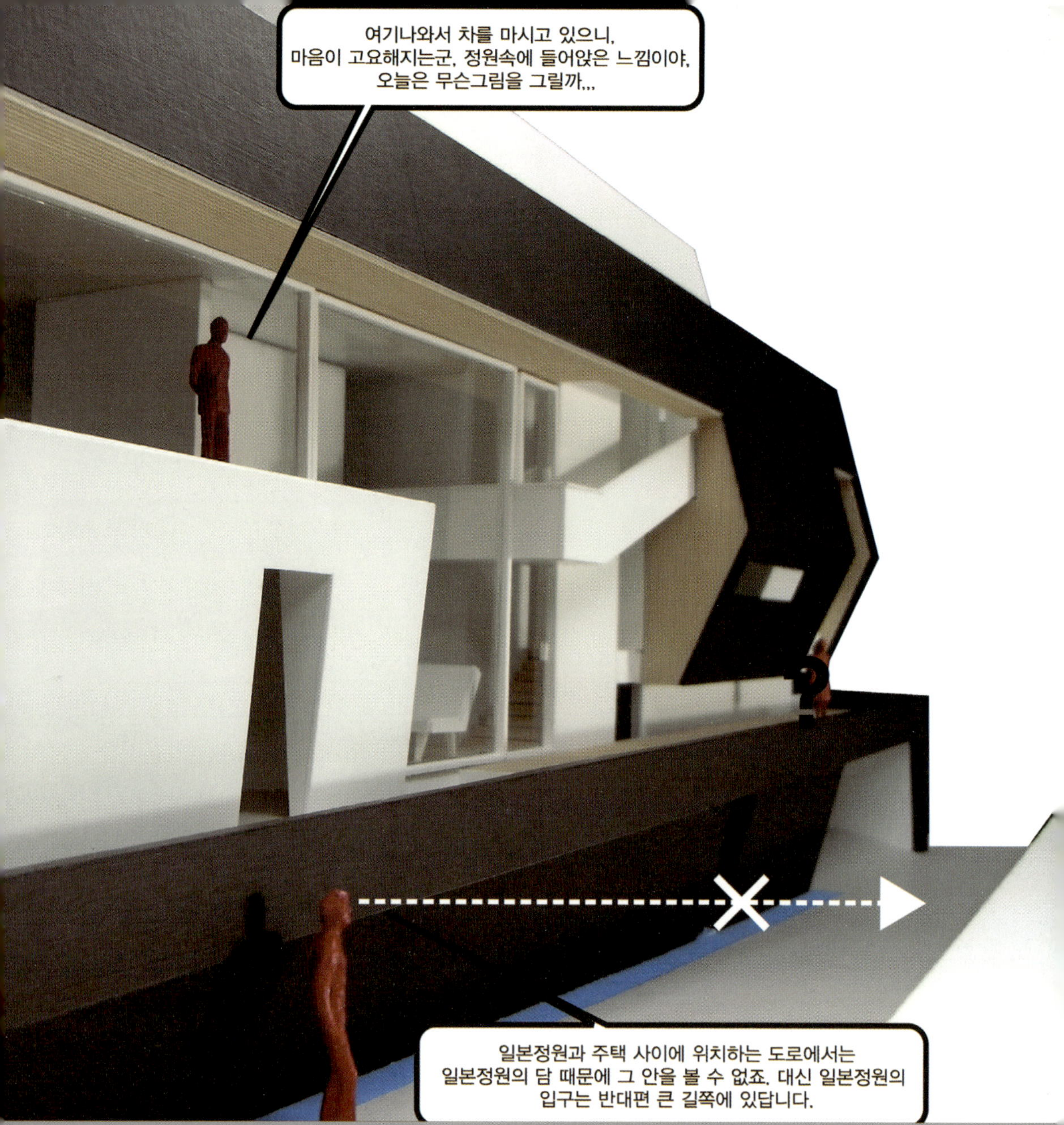

사적인 접근은 높은 담에서, 공적인 접근은 경사진 도로를 따라 낮은 담에서, 각 대지로 두 개의 접근로를 만들고, 단면상 다양한 레벨에서의 조망을 조직하여 방문객이 거주자의 사적영역을 침범하지 않고 아름다운 일본정원을 즐길 수 있도록 하였다.

While, we made dual access for each site, private one at a high wall and public one at a low wall along the sloping road, and organized lines of sight through multi-leveled sections so that guests could enjoy the beutifual Japanese garden without invading the residents' privacy.

요약해서 말하면, 개인주택은 실제로는 공적이지만 가상적으로는 사적인 정원을 가지고, 건축 그 자체로 다른 공간과 분리없이 도시 경계와 엮여있는 전이공간이다. 그래서 우리는 이 주택을 연결과 동시에 분리된 주택(attached detached house)이라고 부를 수 있을 것이다.

Briefly speaking, this detached house has a garden which is actually public but virtually private, and the architecture itself is a transitional space that interlocks urban bundaries without detaching itself from the other.
Thus we may call it an attached detached house.

TOWARDS SHIN CHUN(STREAM)

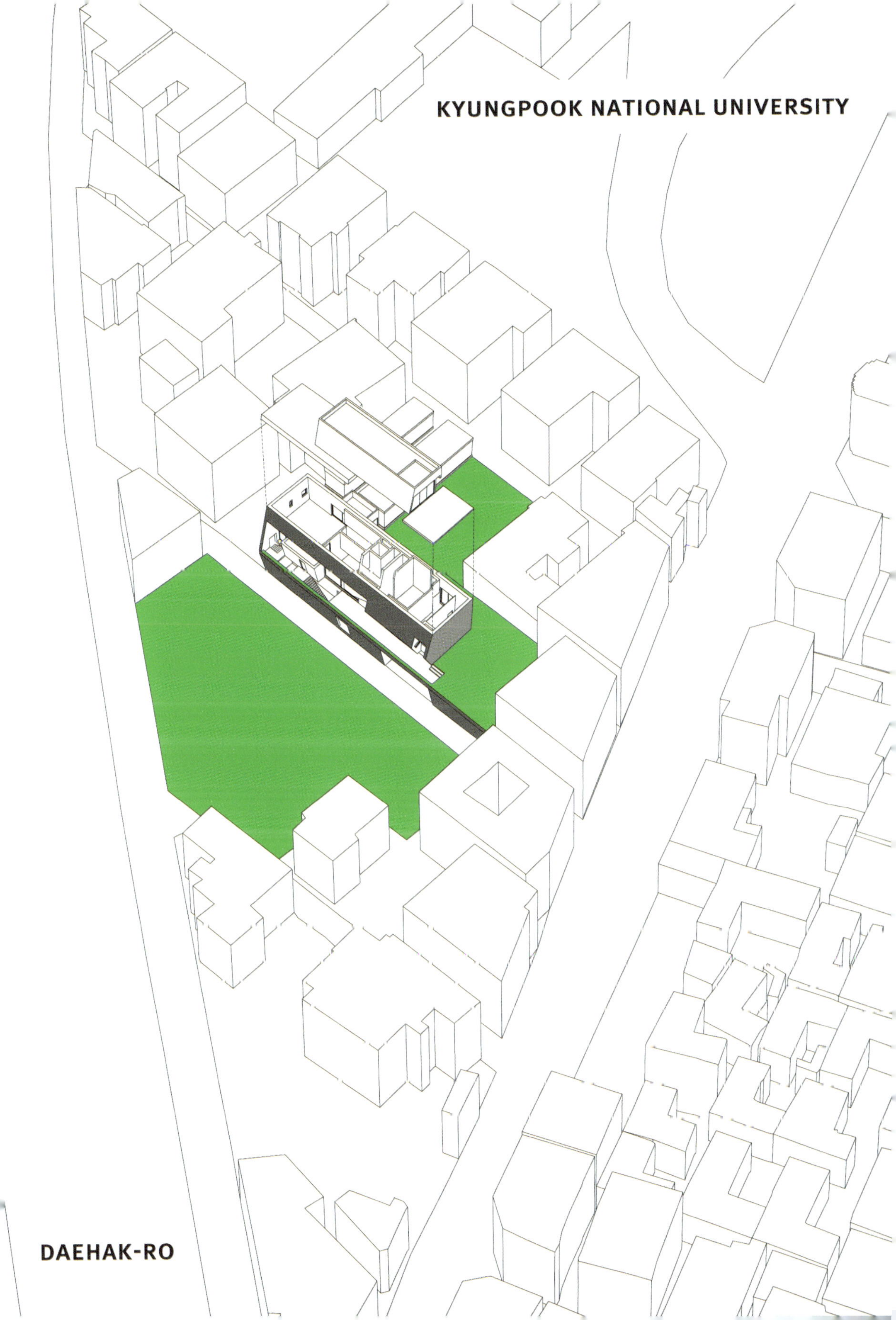
KYUNGPOOK NATIONAL UNIVERSITY
DAEHAK-RO

By doing so, we come across an opportunity beyond efficiency, leading to an innovative collective form which embraces specific individual interests without sacrificing public life.

With this design process, we construct a freed up space for sharing, which embodies a life of encounters and interactions, and help people to form their own public life by themselves. This is what politics for. And this is what we try to achieve, as an architect, who is a politician without party...

매끄러운 형상

이 공동주택은 단독주택과 건축물 사이의 중간 규모에 속한다. 대지는 700년된 녹나무가 승강장과 캐노피를 가로지르는 작은 철도역의 서쪽편에 위치해 있다. 다양한 환경요소(자전거 주차장, 신사, 고가 승강장 하부에 위치한 상점, 소형 스트릿 스탠드 등)가 혼합되어 있는 서쪽편은 소규모 철도역 광장과 로터리로 개발중인 동쪽편보다 혼잡하고 활기있다. 5개의 경계선을 갖는 이 대지는 너무나 많은 요소들로 둘러쌓여 있다. 그러나, 1층 오피스 건물이 면해 있는 서쪽편 전면도로와 북쪽으로 보이는 고가 승강장, 그리고 남측으로부터의 햇빛은 특히나 중요한 것들이었기에, 이 3개의 요소들을 부드럽게 연결하는 건물을 생각하게 되었다.

형태와 환경

일반적으로 일본에서 1인 가족을 위한 공동주택은 약 20 내지 $30m^2$ 규모다. 평면은 4m x 8m 모듈이며 욕실, 화장실, 복도, 그리고 현관이 이 면적의 대부분을 차지한다. 따라서 유효 거실면적은 매우 작다(그림1). 또한 단위세대의 변화는 단지 벽지를 바꾸는 것에 그치고 있다. 만일 단위세대를 2m x 16m로 만든다면, 이 평면은 한쪽면이 길어져서 마치 복도와 같은 방이 된다(그림2).

욕실과 침실은 끝에 배치되고, 방의 형태를 구부림으로써 벽을 세우지 않고도 사적인 영역은 감추어진다(그림3).

전체 볼륨은 한 층에 2개의 단위세대씩 3개층으로 구성된다. 대지경계를 따라 2m 폭의 벽처럼 생긴 볼륨을 배치하고 외부 계단의 위치를 조절함으로써 이 공간은 그 자체의 형태에 의한 단위세대의 변화방식과 환경과의 관계를 생성해낸다(그림 4). 이러한 프로세스를 통해 만들어진 매우 좁은 마당과 외부계단은 주변 건물의 배면으로 둘러쌓인 일반적인 공동주택과 비교해 볼 때 매우 안락한 공간이다. 창은 마당의 채광과 환기를 위해서 2m폭으로, 짝(방의 중심선에 대칭이 되도록)을 이루어 정렬되었다.

익명적인 평면들의 상상력

많은 사람들이 아파트를 찾는 몇 가지 경험을 갖게될 것이다. 우리는 때때로 주택시장에서 다양한 평면들과 마주치게된다. 몇몇 아파트들은 이상하리만치 커다란 발코니를 가지고 있거나 미로 같은 모양을 하고 있고, 매우 긴 형태를 갖는다. 익명적인 평면들의 이와 같이 매력적인 특성들은 단지 많은 건축가들만이 찾는 것들이다. 따라서 나는 익명적인 평면들의 상상력을 디자인 방법으로 이용하길 시도하였다. 나아가, 단지 이와 같은 평면들과 체크리스트들만이 이 프로젝트와 같은 임대주택 부동산 시장에 배포된다. (남쪽으로 면한 방?, 건물의 모서리에 배치?, 발코니의 유무?, 냉난방 장치?, 욕실과 분리된 화장실?)(그림 5).

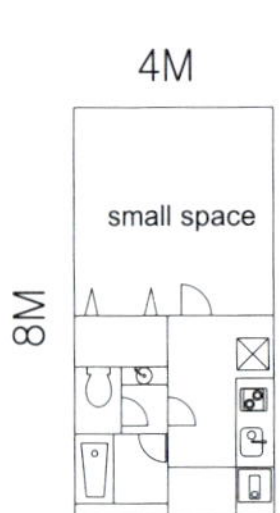

fig.1

fig.2

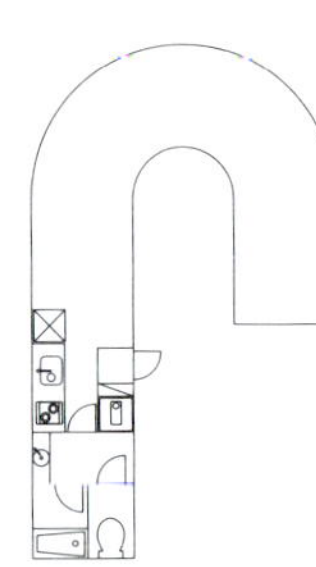

fig.3

각각의 방들은 가능한 많은 검색조건들에 부합되도록 디자인되었고, 각각의 기능적인 방의 형태가 공간구성의 독특한 성격 때문에 어마어마한 양의 분류된 항목에 묻히지 않는다(그림 6). 건축적 구성을 이용한 검색 엔진의 최적화는 부동산 시장에서의 건물의 새로움 대신에 오랜 기간에 걸친 상업적 가치를 만들 수 있다.

Smooth configuration

This collective housing is an intermediate scale between detached houses and buildings. This The site is situated at located in the west side of a small station, where is in a big 700-years-old camphor tree, passing through a plat form the platform and a canopy.

The west side where is being mixed multiple environmental elements (bicycle-parking spaces, a shrine, shops under elevated platforms, small street stands, etc) is more congested and lively alive than the east side being developed a small station station's square and a roundabout in order. This site having 5 borders is enclosed with too many elements, but the front road on the west side the 1st floor office adjoining, the elevated platform being seen to the north, and sunlight to the south are particularly important things. So, I thought it is considered to design the building connecting these 3 elements smoothly.

Shape and environment

Ordinarily collective housing for single family is 20-30m^2 in Japan. The plan is composed of 4m X 8m span, and bathroom, lavatory, corridor and entrance occupy the large part of its area. So, usable an available living space is very small (fig.1). And dwelling variations some variations of dwelling are made by only changing its wallpapers. If a dwelling unit the size of dwelling unit is 2m X 16m, this plan is widen to only one side is able to wide only one side and becomes a corridor like room a room(fig.2).

Bathroom and bedroom are allocated at the end point and privacy area is hidden by bending the room shape, not getting walls up (fig.3). Whole volume is consisted of 2 rooms / 1 floor X 3 layers. Allocating this 2m wide wall-like volume along the border line of lot, and controlling the position of open-air stairs, generate the method of dwelling variations by its shapes and the relation relationship to its environment (fig.4). A very thin courtyard generated by these process and open-air stairs are very comfortable space compared to those of ordinary collective housings enclosed in the back of neighboring buildings. Windows are aligned by a pair (symmetrical to room center line) for ventilating and daylighting the courtyard through 2m volume.

Imaginative power of anonymous plans

Many people will have some experiences in apartment hunting. We often encounter diversified plans in the market. Some apartments have strangely huge balconies, and are labyrinthine, are extremely long. labyrinthine and extremely long. These fascinating aspects of anonymous apartment plans are only discovering things for many architects. So, I made it is made of an attempt to use the imaginative power of anonymous plans as the designing method. Furthermore, only these floor plans and check lists are distributed in the real estate market for rental housings like this project. (rooms facing the south?, allocating at the corner of a building?, with or without balcony?, air-conditioning equipment?, separating toilet from bathroom?) (fig.5).

Each room is designed to correspond with more search conditions as possible, and each functional room shape isn't the shape of each functional room is not buried in enormous amount of classified ads because of its prominent character of spatial composition (fig.6).

Search engine optimization by using the architectural composition can make the long-term commercial value instead of building's freshness in the real estate market.

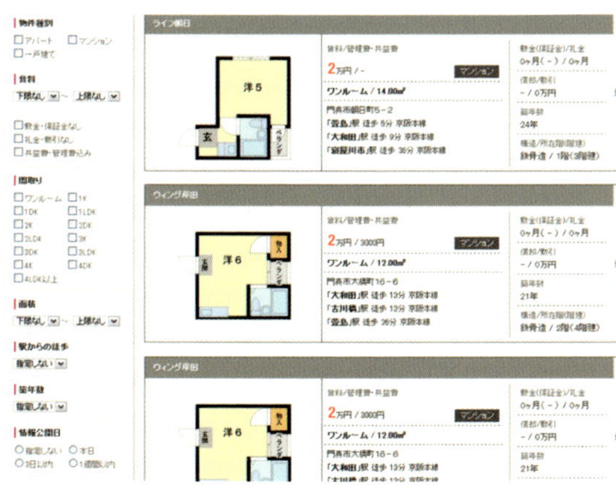

fig.5

fig.4

押入
浴 WC
和6
DK4.5
玄
洋8
ベランダ
物入
玄関
玄
洋7
収納
洋7
玄関
押入
ベランダ
DK5
洋6
CL
ベランダ
玄
洋6
ベランダ
収納
洋7.5
玄
洋20
玄関
洋8.3
K2.9
K2
洋6
物
バルコニー
洋7.5
K
物入
ロフト
洋7
押入
玄関
洋7
押
玄
ベランダ
K2.5
玄関
洋6
押入
K2
和4.5
和6
洋4.5
DK5.5
洗面
収
和6
押入
洋4.5
ベランダ
K2
洋6
玄関
物入
ベランダ
玄
洋6
ベランダ
洋5
玄
ベランダ
押入
洋6
K
ベランダ
玄
物
洋6
玄
K3
玄関
洋6
押
ベランダ
LDK15
玄関
洋4.5
CL
洋7
バルコニー
玄関
K3
洋7
K2
洋8
物入
CL
洗面
CL
洋7
K4
ベランダ

LDK 8
和6
和6
押入
ベランダ
洗面
WC
玄
和6
和6
DK6
押入
押入
ベランダ
洗面
K1.6
洋6
玄関
押
バルコニー
洋7
玄関
ベランダ
LDK15
洋6
押
バルコニー
玄
LDK10.5
物入
バルコニー
玄関
CL
洋3.2
和6
押入
洗面所
玄関
DK6
洋6
ベランダ
物入
MB
洋3
洗
玄
ロフト5
洋20
玄関
和6
K
玄関
押入
洋7.1
DK5
押入
ベランダ
CL
玄関
LDK10
和6
和6
洋6
押入
収
玄関
バルコニー
押入
和6
K2
ベランダ
玄
和6
和6
LDK 11
洋4.5
押入
押
玄関
押入
押入
洋4.5
和6
庭
洋8
玄
洋6
K
押入
玄関
UB
ベランダ
洋20
玄関
洗面所
CL
洋6
K2
玄
バルコニー
洋6
押入
玄
ベランダ
LDK9.5
和6
和6
押入
玄関
バルコニー
洋8
玄関
下駄箱
CL
CL
LDK10.4
収
バルコニー
K3
CL
洋3.8
玄関
MB
室8帖
洋6
収
DK 4.9
収
洗面
バルコニー
LDK12
和6
洋6
物入
押入
玄関
バルコニー
K3
和4.5
和6
玄
押入
ベランダ
玄関
K
和6
押入
DK 4.9
洋6
収
洗面
バルコニー
玄関

Privacy area is hidden by bending the room shape

HEAVEN'S
BAR

モリケン

Aerial view from east

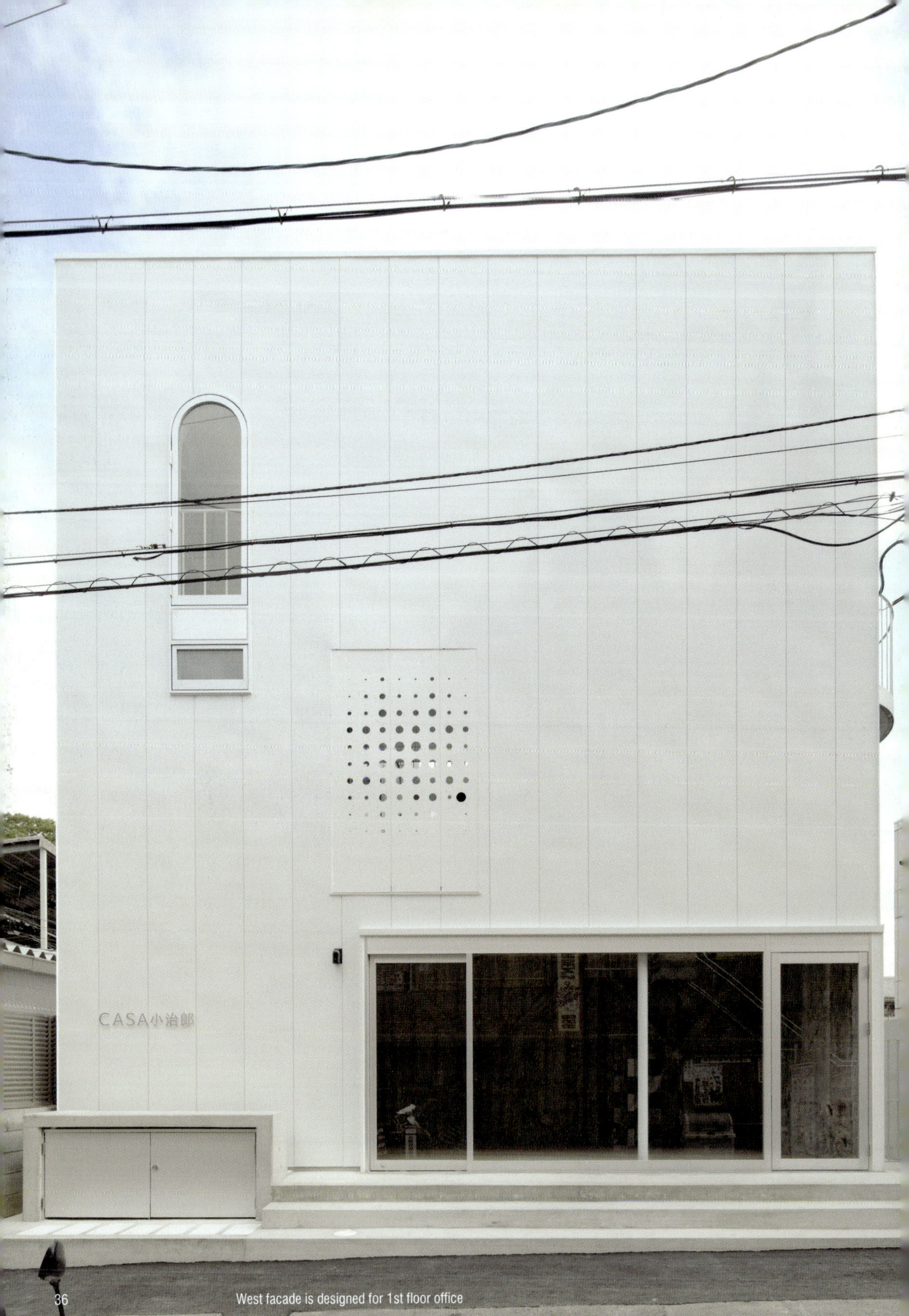

West facade is designed for 1st floor office

Approach to each room

Downward view of courtyard from room 201 terrace

Downward view of courtyard from room 201 terrace

Room 201

Room 201

Room 202. Terrace is shut out by the perforated panel.

Room 301. View from entrance

Room 301. Sunlight is passing through 2m volume

Room 302

Room 302

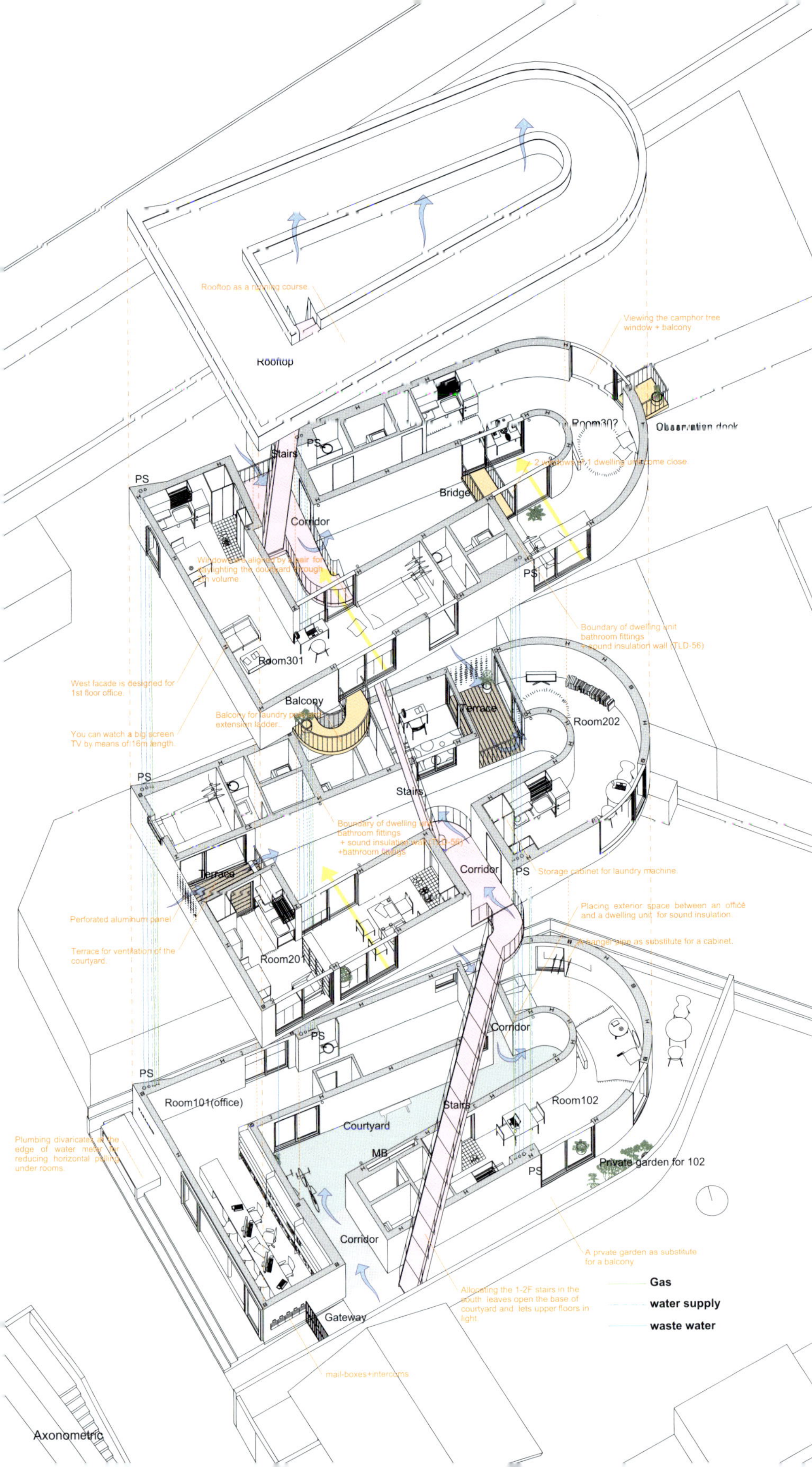

Rooftop as a running course.
Rooftop
Viewing the camphor tree
window + balcony
Room302
Observation dock
2 windows of 1 dwelling unit come close
Stairs
PS
Bridge
Corridor
Windows are aligned by a pair for
daylighting the courtyard through
3th volume.
Boundary of dwelling unit
bathroom fittings
+ sound insulation wall (TLD-56)
Room301
West facade is designed for
1st floor office.
Balcony
Balcony for laundry pole with
extension ladder.
You can watch a big screen
TV by means of 16m length.
Terrace
Room202
Boundary of dwelling unit
bathroom fittings
+ sound insulation wall (TLD-56)
+bathroom fittings
Corridor
Storage cabinet for laundry machine.
Placing exterior space between an office
and a dwelling unit for sound insulation.
A hanger pipe as substitute for a cabinet.
Perforated aluminum panel
Terrace for ventilation of the
courtyard.
Room201
Room101(office)
Room102
Courtyard
MB
Plumbing divaricates at the
edge of water meter for
reducing horizontal piping
under rooms.
Private garden for 102
Corridor
A private garden as substitute
for a balcony.
Allocating the 1-2F stairs in the
south leaves open the base of
courtyard and lets upper floors in
light.
Gas
water supply
waste water
Gateway
mail-boxes+intercoms
Axonometric

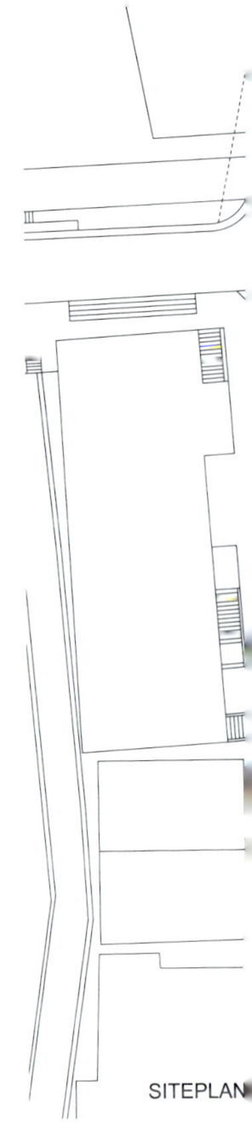

SITEPLAN

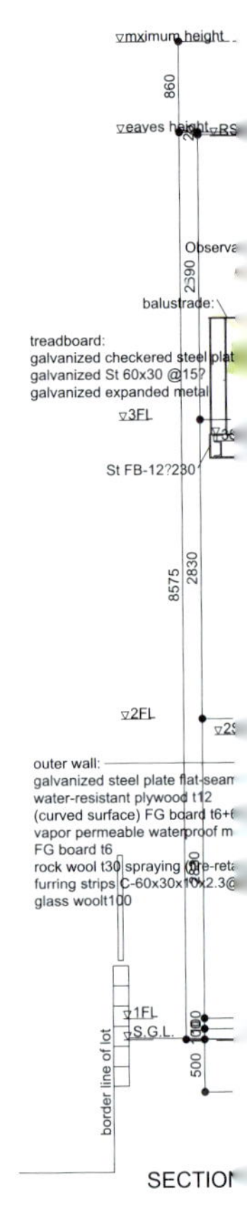

maximum height
860
eaves height
2390
balustrade:
treadboard:
galvanized checkered steel plat
galvanized St 60x30 @15?
galvanized expanded metal
3FL
St FB-12?230
8575
2830
2FL
outer wall:
galvanized steel plate flat-seam
water-resistant plywood t12
(curved surface) FG board t6+
vapor permeable waterproof m
FG board t6
rock wool t30 spraying
furring strips C-60x30x
glass woolt100
1FL
S.G.L.
500
border line of lot
SECTION

Platform
Camphor tree
Exit
Convinience store
Shrine
View axis from platform
Shrine gate
View of Camphor trees
Bicycle-parking space
Sunlight
Corridors of water

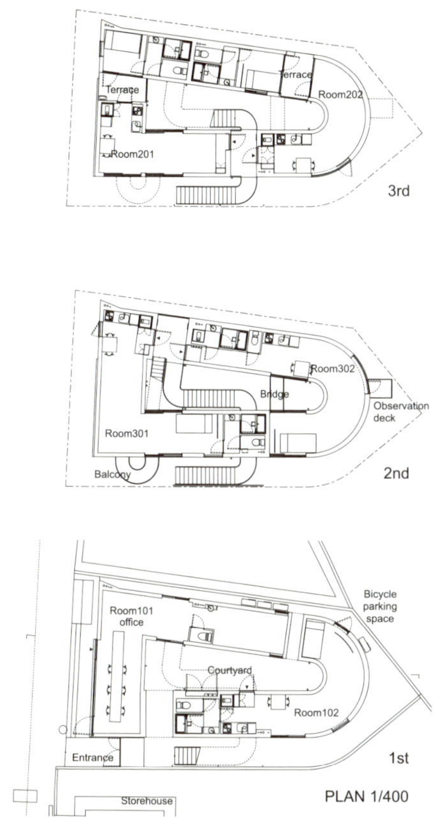
Terrace
Terrace
Room202
Room201
3rd
Room302
Bridge
Observation deck
Room301
Balcony
2nd
Room101 office
Bicycle parking space
Courtyard
Room102
Entrance
1st
Storehouse

PLAN 1/400

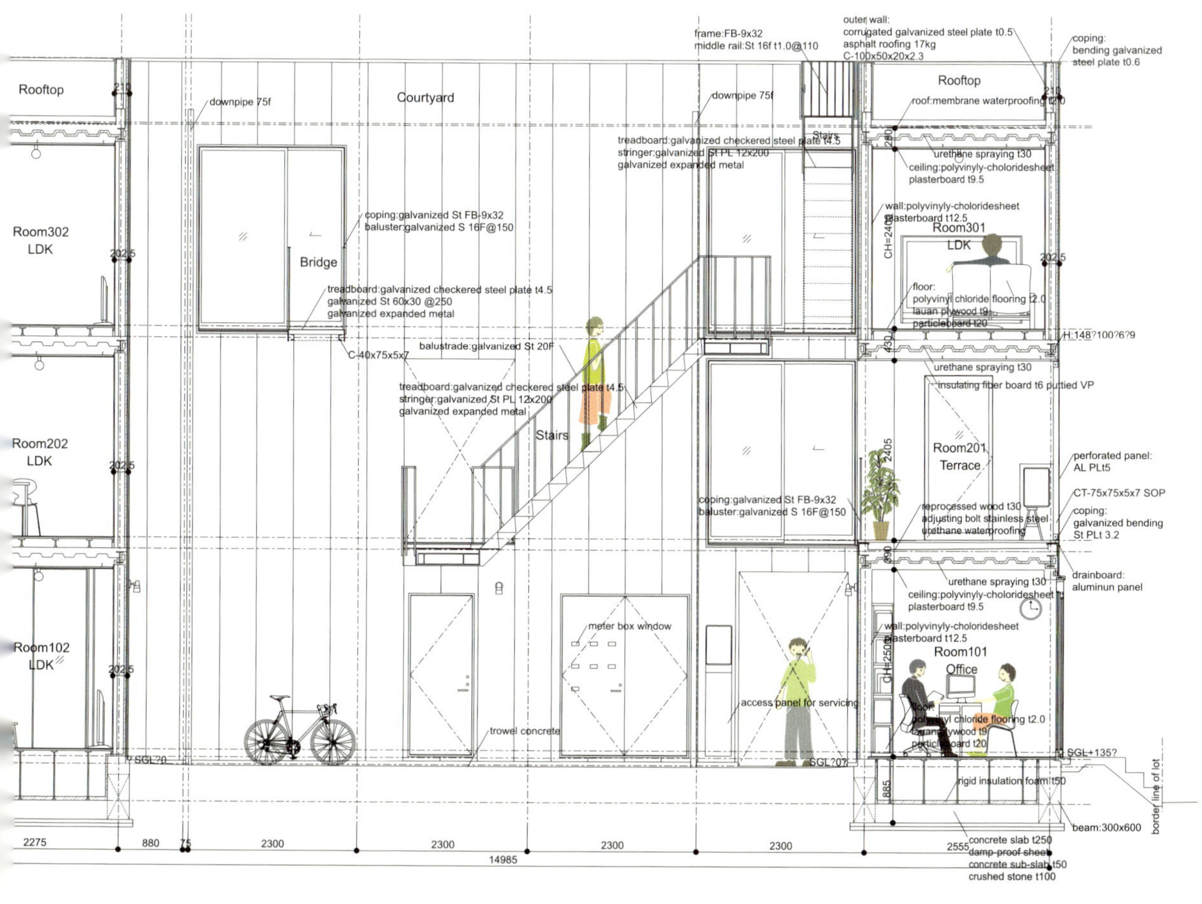
frame:FB-9x32
middle rail:St 16f t1.0@110
outer wall:
corrugated galvanized steel plate t0.5
asphalt roofing 17kg
C-100x50x20x2.3
coping:
bending galvanized
steel plate t0.6
Rooftop
Rooftop
downpipe 75f
Courtyard
downpipe 75f
roof:membrane waterproofing t20
treadboard:galvanized checkered steel plate t4.5
stringer:galvanized St PL 12x200
galvanized expanded metal
Stairs
urethane spraying t30
ceiling:polyvinyly-cholorideshee
plasterboard t9.5
Room302
LDK
coping:galvanized St FB-9x32
baluster:galvanized S 16F@150
Bridge
wall:polyvinyly-cholorideshee
plasterboard t12.5
Room301
LDK
CH=2400
treadboard:galvanized checkered steel plate t4.5
galvanized St 60x30 @250
galvanized expanded metal
floor:
polyvinyl chloride flooring t2.0
lauan plywood t9
particleboard t20
H:148?100?6?9
balustrade:galvanized St 20F
C-40x75x5x7
urethane spraying t30
insulating fiber board t6 puttied VP
treadboard:galvanized checkered steel plate t4.5
stringer:galvanized St PL 12x200
galvanized expanded metal
Stairs
Room202
LDK
Room201
Terrace
perforated panel:
AL PLt5
CT-75x75x5x7 SOP
coping:galvanized St FB-9x32
baluster:galvanized S 16F@150
reprocessed wood t30
adjusting bolt stainless steel
urethane waterproofing
coping:
galvanized bending
St PLt 3.2
drainboard:
aluminun panel
urethane spraying t30
ceiling:polyvinyly-cholorideshee
plasterboard t9.5
meter box window
wall:polyvinyly-cholorideshee
plasterboard t12.5
Room102
LDK
Room101
Office
CH=2500
access panel for servicing
trowel concrete
floor:
polyvinyl chloride flooring t2.0
lauan plywood t9
particleboard t20
SGL-70
SGL?0?
SGL+135?
rigid insulation foam t50
border line of lot
beam:300x600
concrete slab t250
damp-proof sheet
concrete sub-slab t50
crushed stone t100
2275
880
75
2300
2300
2300
2300
2555
14985
885
210
210
2025
2025
2025
2405
430
280
390

D-APARTMENT (casa kojiro)

location: kadoma-city, Osaka
principal use: Collective housing, office
family composition: 5 single family

design
design: SPACESPACE
Takanori Kagawa, Junko Kishigami
stractural design: OHNOJAPAN
Hiroshi Ohno
general constractors
Panahome Co.,Ltd. Yuji Higashihara
Kouji Nakata, Naohiro Hashimoto
construction of structure
Touken Co.,Ltd. Toshio Terasawa
Photo: Koichi Torimura

area
site area: 161.72m²
building area: 94.96m²
total floor area: 228.33m² all(240.16m²)
1stfloor 75.27m² 2ndfloor 76.77m²
3rdfloor 76.29m2
building-to-land ratio: 58.72% (80%)
floor-area ratio: 141.19% (200%)
number of stories: +3

dimensions
maximum height: 9,444mm
upper eaves edge height: 8,584mm
story height: 1st: 2,830mm
2nd: 2,830mm / 3rd: 2,715mm
ceiling height: 1st: 2,400-2,500mm
2nd: 2,400mm / 3rd: 2,400mm

site conditions
district: neighborhood commercial district
fire-proofing district
road width: west2.48-2.45m
parking lot: 0

structure
structure: steel
basis: independent footing

equipment
air-conditioning:
all-in-one air-conditioning system,
thermal source: gas, sunlight

process of works
design period: 2010/6-2010/12
construction period: 2011/6-2011/12

outside finishing
roof: membrane waterproofing t2
outer wall: galvanized steel plate t0.4
window: aluminum,steel sash
exterior: trowel concrete

inside finishing
[LDK,bedroom]
floor: polyvinyl chloride flooring t2.0
wall: polyvinyly-choloride sheet
+plasterboard t12.5
ceiling: polyvinyly-choloride sheet
+plasterboard t9.5
[lavatory,toilet]
floor: polyvinyl chloride flooring t2.0
wall: polyvinyly-choloride sheet
+plasterboard(WP) t12.5
ceiling: polyvinyly-choloride sheet
+plasterboard(WP) t9.5
[terrace,corridor]
floor: recycled wood t30
checkered steel plate t4.5
ceiling: insulating fiber board t6 VP painting

used equipment
air conditioning: Panasonic
sanitary: TOTO AICA Kaijyuuken
kitchen: Mitsubishi Toyoura TOTO
lighting: Saita, DAIKO, Panasonic

rent, unit area
housing 5 + office 1
housing unit area: 33.46m2-38.22m2
rent: 59,000-64,000yen

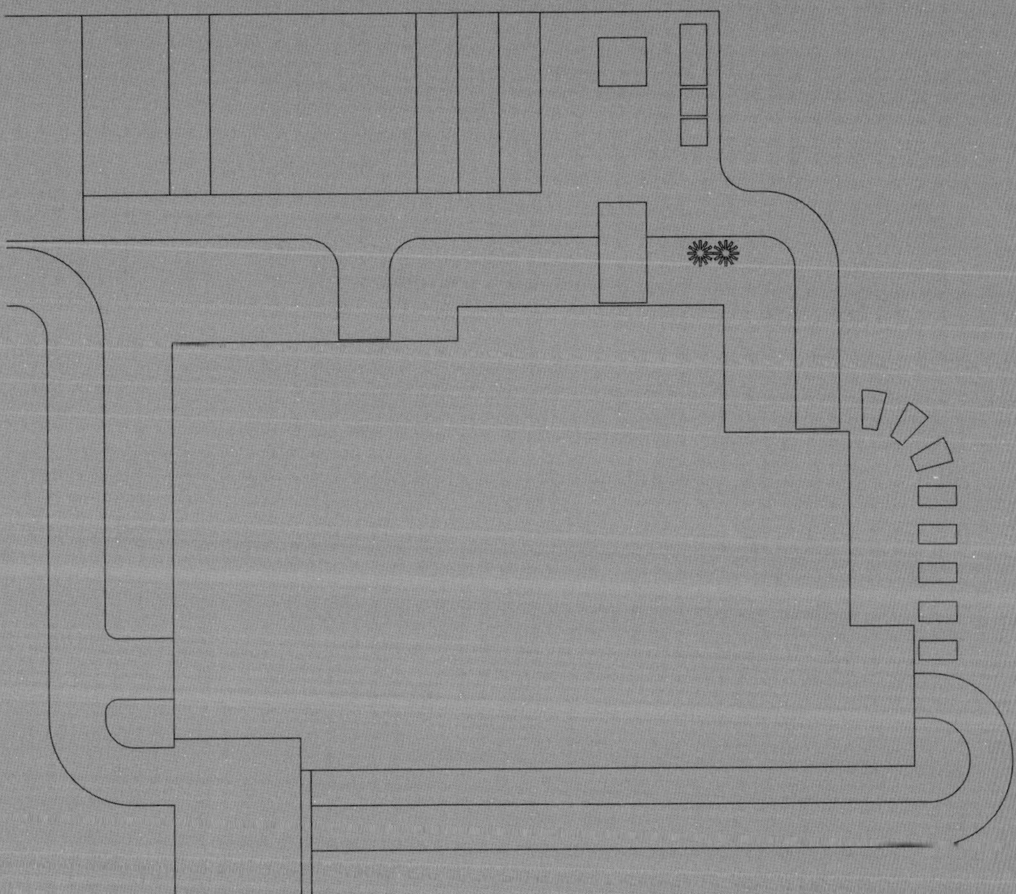

RE-FORMED GARDEN HOUSE

1201

location:	Ikoma-city, Nara
principal use:	residence
site area:	211.58m^2
building area:	83.14m^2
total floor area:	124.39m^2
structure:	LGS
number of stories:	+2
completion:	January, 2012

전형적인 교외 주거지역

대지는 오사카 도심보다 겨울에 조금 더 추운, 나라 현 이코마 시(市)의 주거지역에 위치해 있다. (이 지역은 일본의 에너지절약기준을 위한 분류체계에 있어서 범주 3에 해당한다. 한편, 오사카와 도쿄는 범주 4에 속한다. 범주 3에는 추운 날씨를 갖는 몇몇 지역이 포함되어 있다.) 하지만, 이 지역은 오사카 경계에 면해있고 거주민들이 통근하는데 불과 30분 정도밖에 걸리지 않아, 오사카와 나라의 교외 주택지로 기능하고 있다.

이 지역은 일본의 전형적인 교외 주거지역이다(그림 1). 내부 도로는 넓고, 주택의 지붕경사는 충분하고, 각각의 주택은 남쪽에 면한 정원을 가지고 있다. 그러나, 도시 전체가 짧은 기간에 개발되어서 유사한 몇 가지 주택의 유형만으로 채워지고 말았다. (실제로, 각각의 주택은 조금씩 다른 디자인인 듯 싶지만, 유사한 법칙과 동시적 동질성을 지니고 있다.)

이러한 주택 유형들은 LGS(경량형강) 공법으로 건설되었고, 자체 부품 개발을 통해서 공사비를 절감한다. 그러나 이와 같이 분리된 공법은 특유한 건축사용승인 신청이 필요하기 때문에, 목구조처럼 공간의 볼륨이나 기둥의 위치를 변경하는 등의 유연성을 확보할 수 없다.

형상화된 정원

젊은 부부가 이 지역의 이와 같은 주택들 가운데 한 채를 선택해서, 주택의 외관 변경을 포함한 진면직인 레노베이션을 요구했다.

레노베이션 전에는 남쪽에 면한 정원의 절반 이상이 불법적으로 지붕을 씌운 주차장으로 사용되었으며, 나머지 공간은 폐허처럼 방치되어 있었다(그림 2).

주택의 외관은 건물의 본채 이외에도, 수많은 외부 요소들과 담장, 내문, 긴이 치고, 지전거 주차장, 조경식재 등으로 구성된다. 이러한 요소들은 도시를 향해 몇 가지 중요한 메시지를 지니고 있으며 커뮤니케이션 장치로 기능한다.

우리는 정원과 주차장을 같은 위치에 쌓아올렸다. 즉, 대지 남쪽에 노출 콘크리트 슬라브의 일부를 들어올린후 그 밑에 주차장을 계획하였다. 이렇게 형상화된 정원은 각각 독립적인 통로를 통해서 부엌과 거실에 연결되며, 외부 거실로 사용될 수 있다. 이밖에도, 이웃과의 더 많은 커뮤니케이션을 만들어내고자 몇 개의 통로들을 대지 경계를 따라 배치하였다(그림 3).

작은 방들로 구성된 1층은 커다란 원룸으로 변환되었으며, 몇몇 LGS 기둥들은 이와 같은 변환과정 속에서 부가적인 기능을 갖는 가구로 탈바꿈되었다(선반, 옷걸이, 조명, 시계, 술자). 마찬가지로 몇 개의 창문들도 벤치나 책상과 같은 가구로 변환되었다(그림 4).

우리는 기존의 외부 구조들과 그것들이 중첩되는 방식을 변화시킴으로써, 주변환경과 연계되고 이웃과 소통하는 새로운 형태를 갖는 주택을 다시-디자인하였다.

fig.1

fig.2

A typical suburban residential area

This site is situated at a residential area in The site is located in a residential area, Ikoma-city, Nara which where is little colder in winter compared to the inner-city of Osaka. (It corresponds to the category 3 area of the classification for energy-saving standard in Japan. Osaka and Tokyo correspond to the category 4. Category 3 includes some cold weather region.) But it shares a border with Osaka and the residents can commute in only 30 minutes, so it functions as a bedroom suburb of Osaka and Nara.

This area is a typical suburban residential area of Japan (fig.1). The inner road is wide, the pitch of each house is a big enough and each house has the southern garden. But the whole town was developed in a short period of time, therefore only some few similar types of house are built on it. (Actually, each house has a different design, but has a similar rule and contemporaneousness.)

These types of houses are built with LGS (light-gauge steel) construction methods. Developing own components lowers their construction costs. But this isolated construction methods uses the peculiar application for building confirmation, so doesn' t it does not have the flexibility to change its volume and the position of columns like wooden constructions.

Formed garden

A young married couple selected one of these houses in this area, and required us the full-scale renovation including to a change the exterior appearance of the house.

Before our renovation, more than half of the southern garden was used as a parking lot with illegal carport, and the remaining fell into ruin (fig.2).

The exterior appearance of houses house is consisted of many exterior elements, such as fence, gate, carport, bicycle parking, plants and so on, not only of its main body. These elements have some important messages to the city, and work as communication communicative devices.

We stacked the garden and the parking lot at the same position. We lifted up the part of an exterior concrete plane configured at the southern part of the site and allocated a parking lot on the downside of its plane. This The formed garden is linked to kitchen and living through each independent passageway. It can be used as an outdoor living. Along other adjacent sites, some passageway was made to generate more communication with neighbors (fig.3). 1st floor is composed of some small rooms was converted to a big one-room. Some appeared LGS columns in this process were furniturized changed to furniture by attached some other functions (shelf, hanger, light, clock and measure). Some The windows were also furniturized functioned as benches and tables (fig.4).

By modifying the existence of exterior structures and the way of their overlapping, we re-designed the house and its appearance exterior with new involvement of its surrounding environment and neighboring communication.

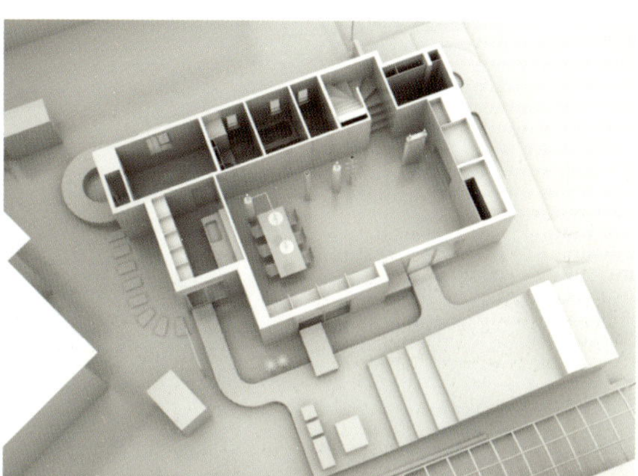

fig.3

fig.4

General view from the south garden.

View of the south garden and parking lot.

View from the front road.

The south garden is connected to kitchen and living room separately.

Rody

Living room.

Furniture columns.

Furniture window.

Living romm and kitchen.

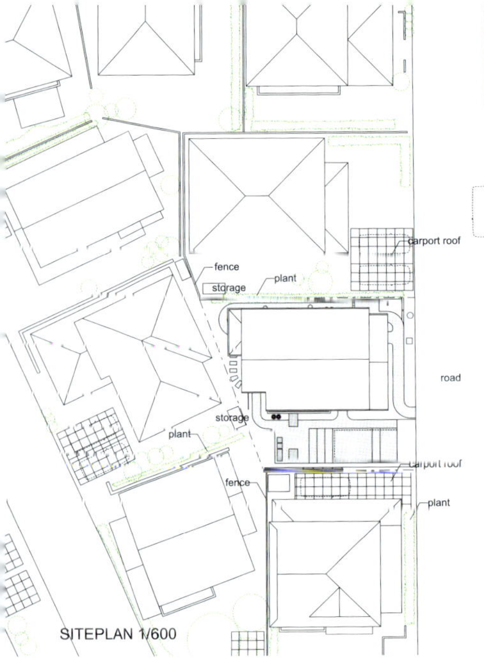

SITEPLAN 1/600

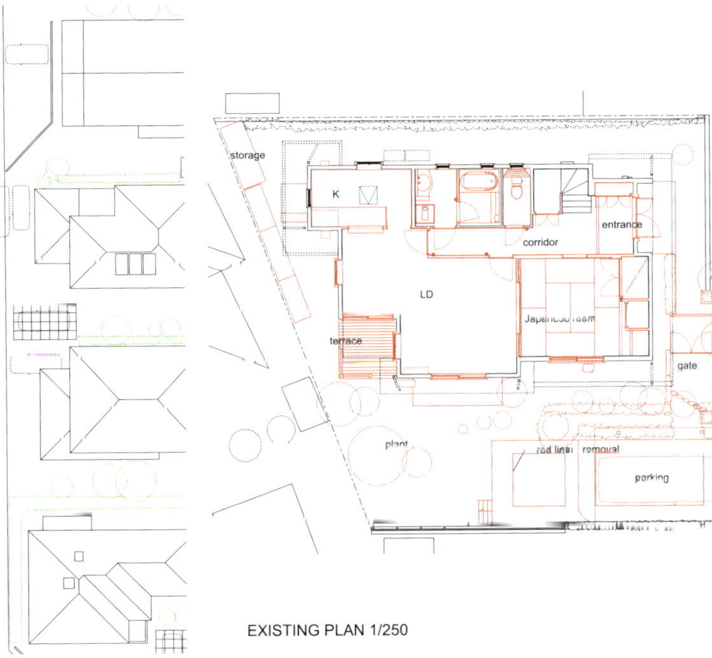

EXISTING PLAN 1/250

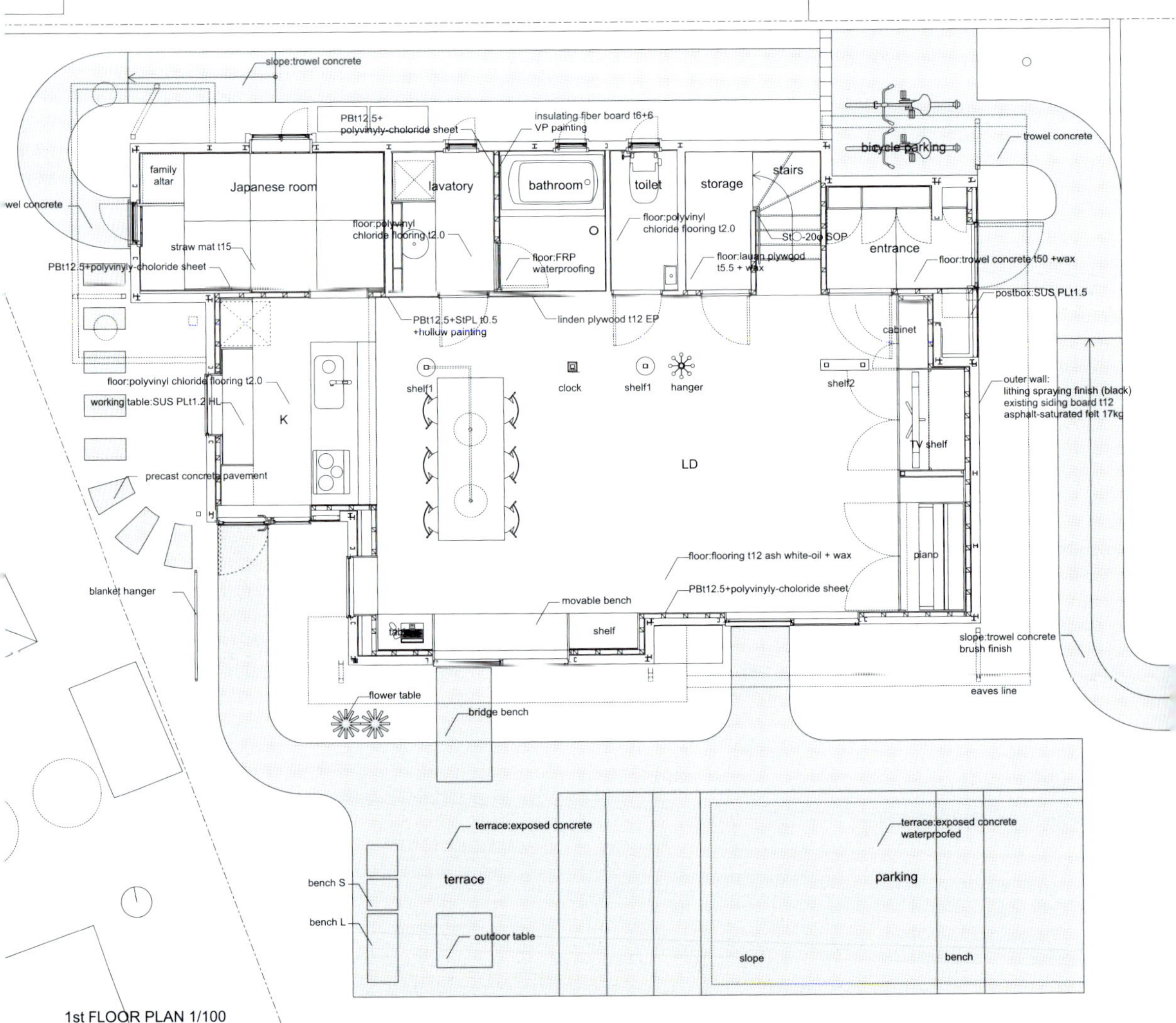

1st FLOOR PLAN 1/100

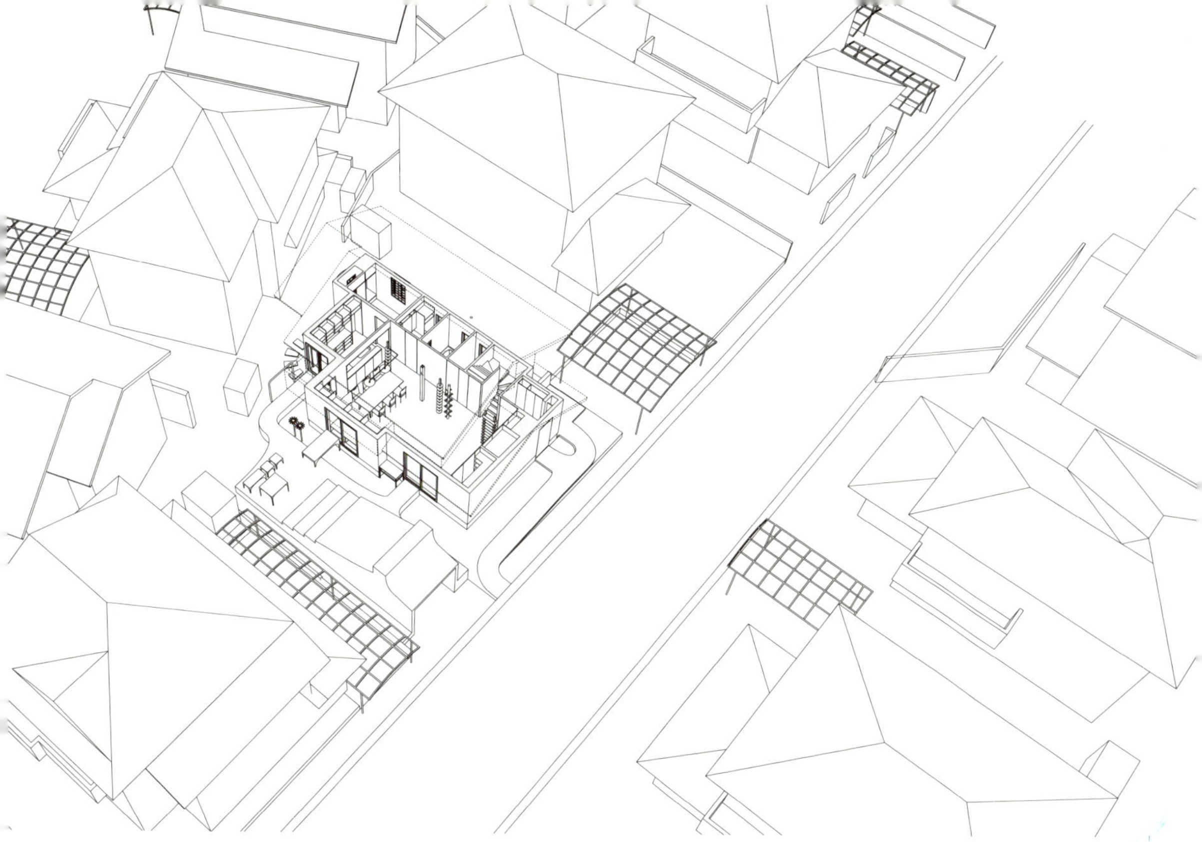

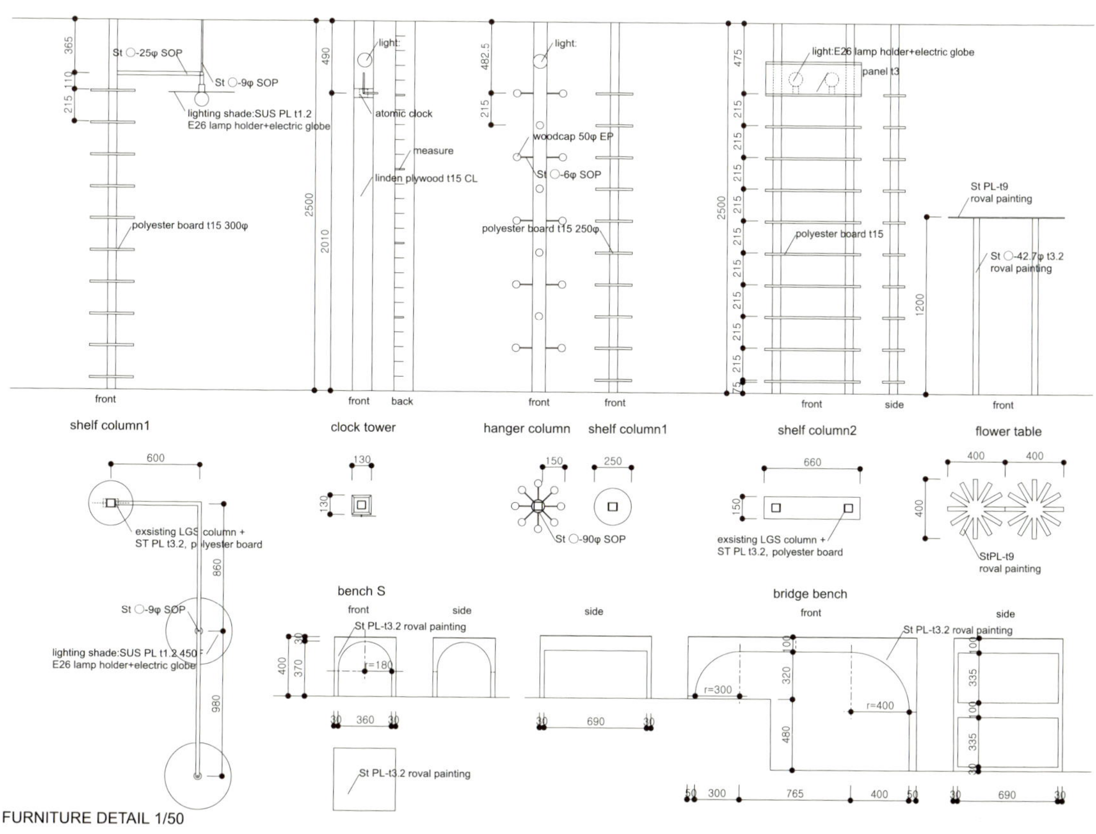

FURNITURE DETAIL 1/50

Re-Formed Garden House

location: Ikoma-city, Nara
principal use: Residence (renovation)
family composition: 2+1
design
design: SPACESPACE
Takanori Kagawa, Junko Kishigami
stractural design: TAPS
Masaichi Taguchi, Shinsaku Kataoka
exterior design: Soji
Takeo Matsushita
general constractors
NEWing Co.,Ltd. Masato Shoji
curtain: fabricscape, Kiyohiko Yamamoto
Photo: Koichi Torimura
area
site area: 211.58m^2
building area: 69.36→83.14m^2
total floor area: 123.14→124.39m^2
1stfloor 68.11→69.36m^2
2ndfloor 55.03m^2
building-to-land ratio: 32.78→39.29% (40%)
floor-area ratio: 58.20→58.79% (60%)
number of stories: +2
dimensions
maximum height: 7,855mm
upper eaves edge height: 5,888mm
story height: 1st: 2,929mm 2nd: 2,422mm
ceiling height: 1st(LDK): 2,500mm
1st(lavatory.toilet.entrance): 2,200mm
2nd: 2,295-2,300mm
site conditions
district: low building residential district
road width: east6.0m
parking lot: 1
structure
structure: LGS
basis: independent footing
equipment
air-conditioning:
all-in-one air-conditioning system,
floor heating
thermal source: gas
process of works
design period: 2011/3-2011/9
construction period: 2011/10-2012/1
outside finishing
roof: existing clay tile roofing + painting
outer wall: existing siding board t12
+ lithing spraying
window: aluminum,steel sash
exterior: exposed concrete
exterior furniture: steel + ROVAL painting
inside finishing
[LD]
floor: flooring t12 ash white-oil + wax
wall: polyvinyly-choloride sheet
+plasterboard t12.5
ceiling: polyvinyly-choloride sheet
+plasterboard t9.5
[K]
floor: polyvinyl chloride flooring t2.0
wall: polyvinyly-choloride sheet
+plasterboard t12.5
ceiling: polyvinyly-choloride sheet
+plasterboard t9.5
[entrance]
floor: flooring t12 ash white-oil + wax
wall: polyvinyly-choloride sheet
+plasterboard t12.5
ceiling: polyvinyly-choloride sheet
+plasterboard t9.5
[Japanese room]
floor: straw mat t15
wall: polyvinyly-choloride sheet
+plasterboard t12.5
ceiling: polyvinyly-choloride sheet
+plasterboard t9.5
[lavatory.toilet]
floor: polyvinyl chloride flooring t2.0
wall: polyvinyly-choloride sheet
+plasterboard t12.5
ceiling: polyvinyly-choloride sheet
+plasterboard t9.5
[bathroom]
floor: FRP waterproofing
wall: insulating fiber board t6+6 VP painting
ceiling: insulating fiber board t12 VP painting
[storage]
floor: lauan plywood t5.5 + wax
wall: lauan plywood t5.5 + wax
used equipment
air conditioning: Mitsubishi
sanitary: TOTO, CERA, Panasonic
kitchen: Mitsubishi, TOTO, Betty&Home
lighting: Saita, DAIKO

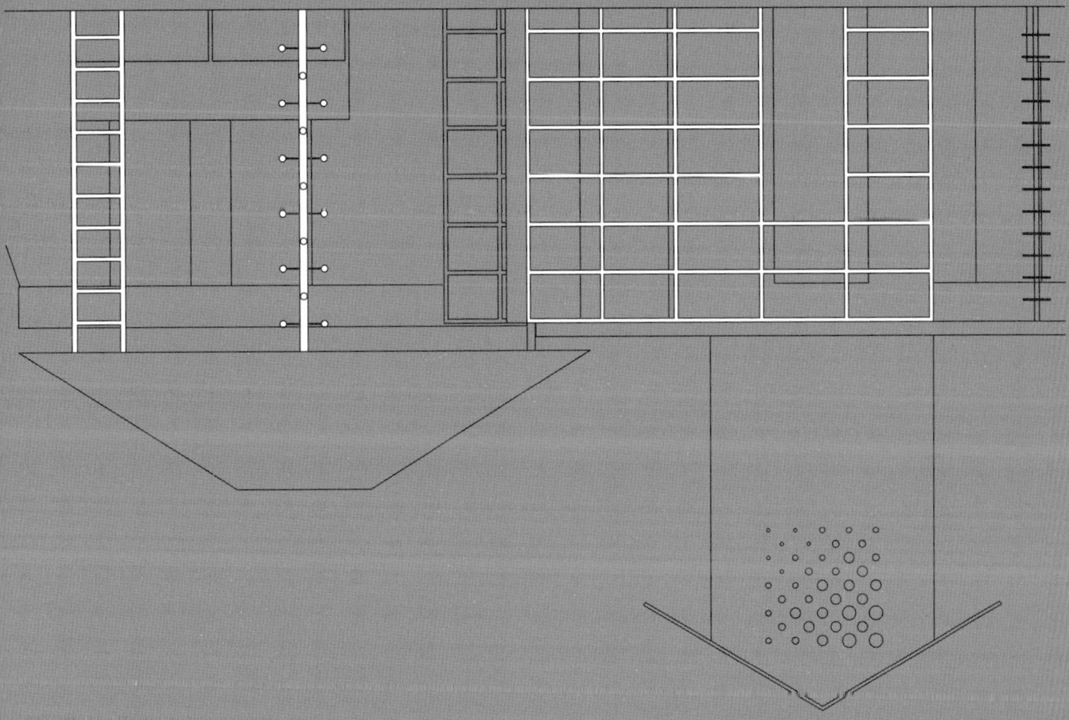

BANKART LIFE 3 - BOOKSHOP

1109

location:	Yokohama, Kanagawa
principal use:	pavilion(residence)
site area:	118.00m^2
building area:	53.28m^2
total floor area:	53.28m^2
structure:	W
number of stories:	+1
completion:	September, 2011

BankART Life3 Shinminatomura (요코하마 트리엔날레)

Shinminatomura(요코하마 디자인전람회의 협동 프로그램의 하나)는 다양한 크리에이터들이 활동하고, 매일 많은 이벤트들이 열리는 작은 도시이다. 주택, 도서관, 공공 홀 등 모든 건축물들은 건축가와 예술가들에 의해 디자인되었다.
Shinminatomura는 4개의 특화된 영역(역, 도심, 학교, 교외)으로 구성된다. 우리는 교외 영역의 작은 집(서점으로 사용될 파빌리온)의 디자인에 참여하였다. 교외 영역은 바다가 조망되는 주거지역으로, 광장 주변으로 주택들과 카페들이 위치한다. 이 집에는 많은 요소와 장면들이 담겨있다. 긴 방(침실과 박공)과 굽은 방(욕실과 화장실)은 이웃한 에어컨 설비를 숨기도록 구성되어 있고, 높은 방(공부방)과 중첩되어 있는 처마에 의해 덮여있는 좁은 골목의 형성을 돕고 있다. 화단과 결합된 외부 계단은 하부로는 개 집(doghouse)으로 사용되고, 위로는 긴 벤치로 사용된다. (그림 1) 사각모임지붕은 거실과 독서공간으로 활용되는 개방된 야외 홀 위에 지어졌다. 이 야외 홀에 있는 테이블과 의자들은 광장을 면하는 외부 가구들과 연계된다.
테이블 위에는 작은 사인(signage)을 가진 테이블 소품들이 배열되어 있다. 책장과 테이블 세트, 그리고 마치 산처럼 생긴 화단을 가진 콘크리트 벤치는 광장에 면해 있다. (그림 2) 세 개의 지붕은 방의 볼륨들과 책 선반들, 그리고 둥근 선반, 사다리, 옷걸이 등과 결합된 기둥들에 의하여 지지된다.(그림 3) 도시는 다양한 건축물뿐만이 아니라, 수많은 도시적 요소들, 극도로 긴 집, 작은 작업장, 노점, 가게 앞의 의자와 테이블, “간판” (のれん, 상점 출입구에 상점이름을 써 넣어 드리운 천 등), 표지판, 우편함, 신호등, 교통 페인트, 가로수, 화분, 화단, 징류장, 지동판매기 등으로 만들어 진다. 이러한 작은 건축물(일반적인 건축물보다 훨씬 더 작은)과 도시의 작은 요소들은 건축가의 디자인을 위한 대상이나 어휘로는 고려되지 않아왔다.
우리는 이러한 요소를 참고하여 발전시킨 대안적 어휘를 통하여 이 집을 디자인하였다.
우리는 이러한 방법으로 도시적 즐거움을 집안에 담을 수 있고, 더욱이 지금까지의 건축이 하지 못했던 다양한 것들을 다루게 하며, 건축을 주변 도시와 재결합시킬 수 있다고 생각한다.

BankART Life3 Shinminatomura (Yokohama Triennale)

Shinminatomura (a cooperative program of Yokohama Triennale) is a small city in which various creators are working and many events are held in everyday. All buildings (houses, library and hall) are designed by architects and artists. Shinminatomura is composed of 4 characteristic zones (station zone, downtown zone, academic zone, suburban zone). We designed a small house (a pavilion used as a bookshop) in suburban zone. Suburban zone It is a residential district overlooking the sea. Houses and a café cafes are located in around a square.
This house has many elements and sequences., such as Long room (bedroom, gable roof) and Bending-room (bathroom and lavatory). They are allocated to conceal neighboring air conditioning equipments, additionally generate a narrow alley being covered by overlapping eaves with Tall-room (study room). Outdoor stairs combined with flowerbeds are used as a doghouse (on the downside) and a long bench (on the reverse side) (fig.1).
A square hipped roof is built over open-air hall used as living room and reading room. Tables and chairs set in open-air hall link linked with outside furniture facing the square.
The top of the tables is arranged by small signage tableware. A small signage tableware is arranged on the top of the tables.
Book stands, table sets and mountain like concrete benches planted flowers are facing to the square (fig.2). Three roofs are supported by room volumes, book shelves and some columns combined with shelf plates, ladder and clothes hangers (fig.3). The city is made of not only various buildings, but also numerous urban elements, extremely long house, small sheds, street stands, storefront benches and tables, "Noren", signboards, mailboxes, traffic lights, traffic paints, boulevard trees, plant pots, flowerbeds, car stops, vending machines and so on. These small buildings (less than ordinary buildings) and urban small elements the small elements of urban have not considered as candidates and vocabularies to design for architects. We designed the house by using alternative vocabularies generated to reference these elements.
We think that this method enables to take the enjoyment of city in the house, furthermore and to deal with various things architecture haven't have not ever done, and to reconnect architecture with surrounding city more.

ig.1

fig.2

STOREFRONT
STOREFRONT
STOREFRONT
ZINE
STOREFRONT
STOREFRONT

Revontulet
drawing
...等

Open living room.

Bending room, Narrow alley.

Narrow alley.

Bending room (lavatory and bathroom) is used as exhibition spac

Interior vew of long room.

Bending room.

1st FLOOR PLAN 1/150

ROOF PLAN 1/150

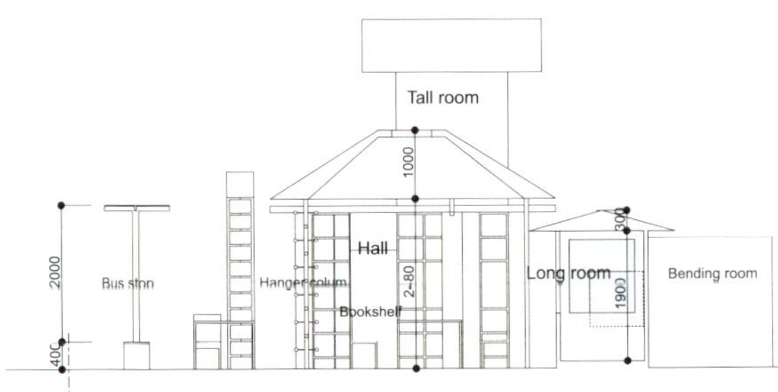

A SECTION 1/150

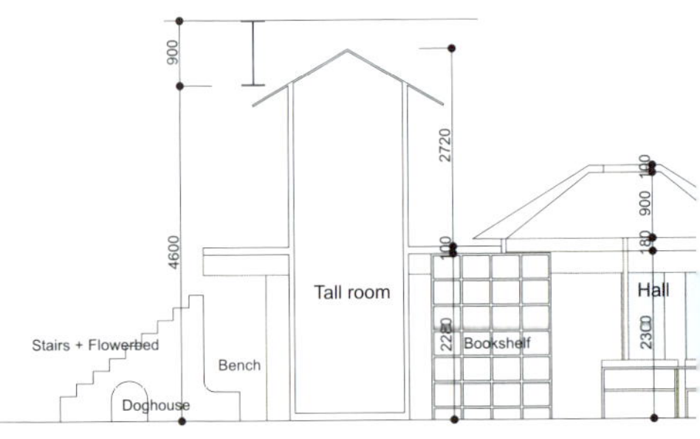

B SECTION 1/150

SITEPLAN 1/800

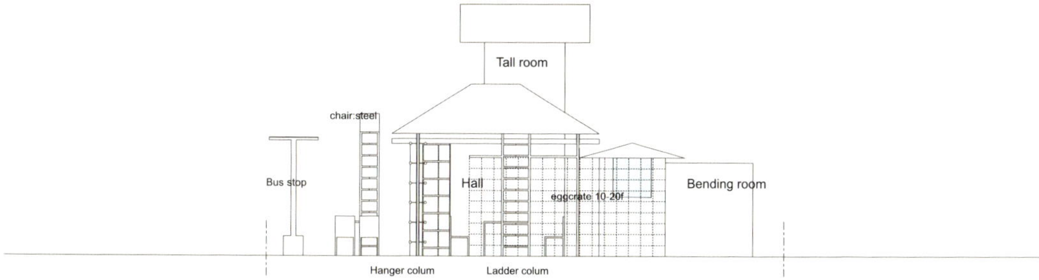

ELEVATION2 1/150

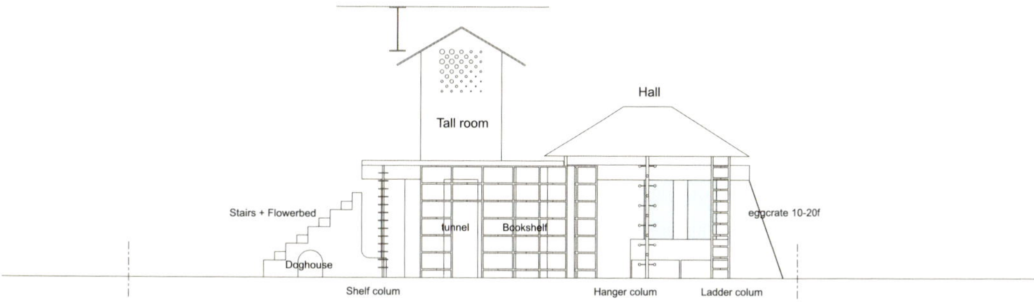

ELEVATION1 1/150

FICTIONAL SITEPLAN 1/500

BankART LIFE3 - Bookshop

location: Yokohama-city, Kanagawa
principal use: Residence (pavillion)

design
design: SPACESPACE
Takanori Kagawa, Junko Kishigami
organizer
BankART 1929
joint hosting
Yokohama, Culture and Tourism Bureau
Photo: Koichi Torimura

area
site area: 118.00m²
building area: 53.28m²
total floor area: 53.28m²
1stfloor 53.28m²
building-to-land ratio: 45.15%
floor-area ratio: 45.15%
number of stories: +1

dimensions
maximum height: 5,100mm
upper eaves edge height: 4,600mm
story height: 1st: 2,480mm
ceiling height: 1st(hall): 2,480-3,380mm
1st(tall room): 4,500-4,900mm
1st(long room): 1,900mm
1st(toilet, bending room): 1,850mm

site conditions
road width: square + 1.8m
parking lot: 0

structure
structure: wood
process of works
design period: 2011/8
term of a session: 2011/8-2011/11

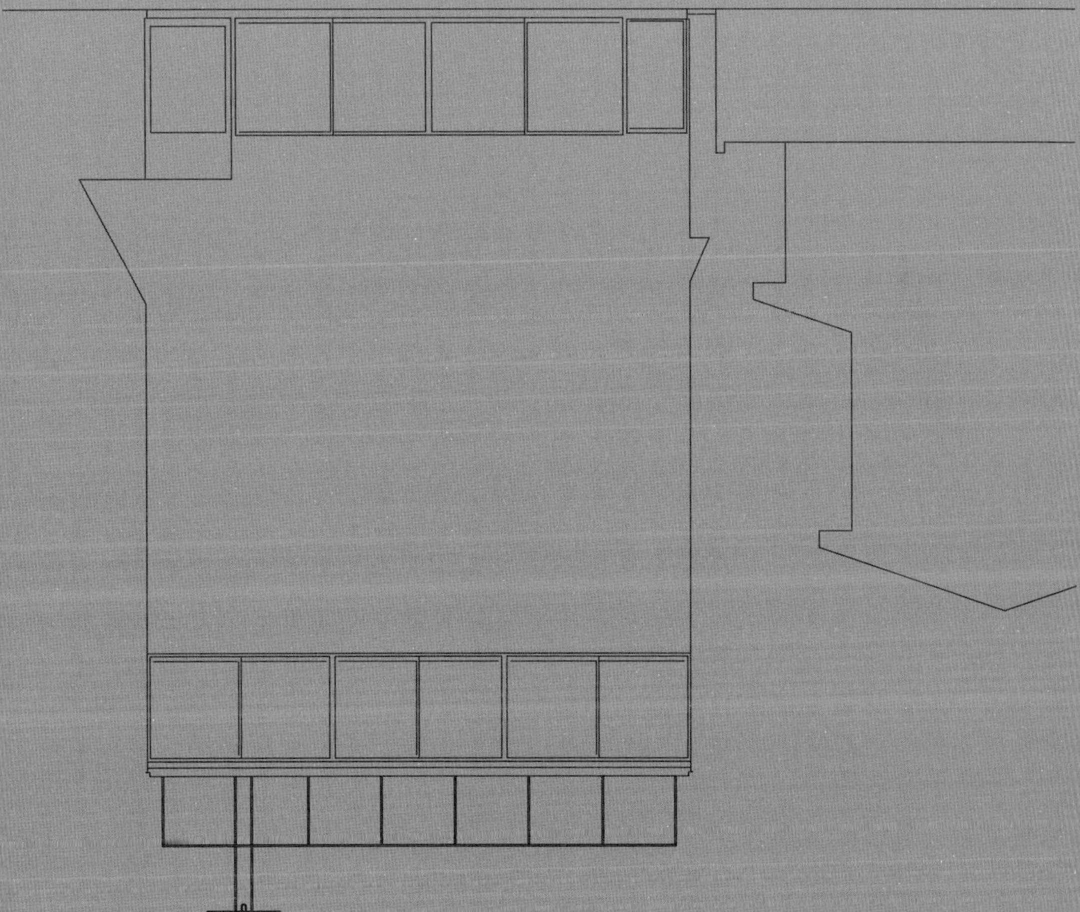

GROUND AND ABOVE ROOF HOUSE

1009

location:	Ibaraki-city, Osaka
principal use:	residence
site area:	53.31m²
building area:	30.96m²
total floor area:	66.92m²
structure:	W
number of stories:	+2
completion:	September, 2010

2층 목조주택

이 주택은 철도역 근처 주거지역 중심가로에서 한 블록 뒤에 있는 길모퉁이에 위치해 있다.
대지는 북쪽과 동쪽으로 2개의 가로에 면해있으며 동쪽(장변)에는 T자형 교차로가 있고 암거구조물 상부 지면에 옥외 주차장이 있다. 이 곳은 특유한 개방감이 있는 장소다.
거리에는 사인보드와 화분들이 놓여져 있고, 이따금 거리 주변에서 만나 잡담을 나누기도 하는 장기 거주민들이 많은 지역이다.

이 대지는 마치 조닝 법규의 변곡점과 같은 특별한 곳이다. (상업지역에서 주거지역으로 바뀌는) 조닝 변화선이 대지의 남측 경계선 근처에 지정되어 있다. 따라서 이 주택은 이 지역의 주변 건물보다 약간 더 높게 지어질 수 있었다. 이와 같은 주변 조건에 따라 2개층의 목조 주택이 건축되었다. 활력있는 외부공간과 연계된 거실을 만들기 위해서 1층은 가로와 연결되었고, 2층은 번화한 거리로부터 물리적인 거리를 두고자 바닥 슬라브를 인접 건물의 처마선보다 더 높이 들어올렸고, 남측으로부터 1층에 충분한 채광을 확보했다.
겨울에는, 내부에 욕실이 있는, 가마쿠라 형태의 언덕이 1층 남측면 상부에 만든 커다란 창을 통해 모아진 태양열을 축열할 수 있다. 이 언덕 모양의 공간은 TV를 볼때는 소파로 사용되고, 창을 열고 바깥에서 걸터 앉을 때는 베란다로 사용되며, 현관에서 들어올때는 스크린으로 사용되며, 창고로도 혹은 아이들의 놀이 공간으로도 사용되는 등, 다양한 용도의 생활용품으로 기능한다.
2층의 침실은 최소 크기로 계획되었기 때문에, 2층의 홀 공간은 공부를 하거나 책을 읽는 등, 잠을 자는 것을 제외한 다른 모든 행위들을 위해 사용된다.
2층 창에서 바라다보는, 지붕 위 풍경은 1층과는 완전히 다른 장면으로 펼쳐진다(그림 1).
주택 내부의 일련의 움직임을 바람과 열, 사용자와 물품들의 움직임과 연계시키는 이와 같은 공간 구성을 현재로서는 "2층 목조주택"이라고 말할 수 밖에 없겠지만, 당장은 이와 같은 구성에 달리 적당한 이름이 없다는 바로 그 까닭에 이 주택은 다른 일반적인 주택과 다르다고 생각한다.

열환경

2층 바닥 슬라브 하부, 1층 남측면 상부에 배치한 커다란 창은 태양광과 태양열을 받아들이기 위해서 디자인된 것으로, 겨울에 이 창을 통해 들어온 햇빛은 콘크리트 언덕에 축열된다. 언덕의 위치와 높이는 한겨울 태양고도와 주변 건물의 처마높이를 고려하여 설계되었다. 축열이 어려운 날을 대비하기 위해서 콘크리트 언덕과 평평한 바닥, 그리고 햇빛이 닿지 않는 공간에는 바닥난방이 설치되었다.
여름에는 전동 블라인드를 통해서 햇빛이 차단되도록 하였고 덥혀진 공기는 창 사이의 공간에 모여져서 2층에 설치된 창으로 빠져나가게 된다. 이를 위해서 남측벽과 북측벽, 그리고 지붕에 2중 벽을 두어 그 사이로 통기 구멍이 설치되었다. 한편, 겨울에는 이 통기 구멍이 축열 공간으로 기능하여 공기흡입구를 통해 따뜻한 공기를 공급한다(그림 2).

언덕 덮개

언덕 덮개는 언덕의 부속 장치로, 콘크리트의 열재생위험과 바닥난방이 늦게 가동되는 문제를 해결하기위해서 디자인되었다. 덮개천은 소파에 사용되는 것을 두겹으로 짜서 만들었으며 쿠션에 우레탄을 끼워 넣어서 축열성능과 내구성을 강화했다.
언덕의 3D 곡면을 만들기 위해서, 입체재단 대신에 평면재단 방식을 이용했으며, 노변과 보넬을 사용하여 디자인한 후, 최종적으로는 샘플 천을 사용하여 각각의 치수를 확정지었다(그림 3). 덮개 윗면의 주머니들은 핸드폰이나 리모콘, 또는 신문을 넣을 수 있도록 디자인되었으며, 더 많은 주머니가 필요할 때도 쉽게 덧붙일 수 있도록 하였다.

A two-story wooden house

This The house is situated at located in a street corner located where is one lot behind from a main street in a residential area near a railway station. The site borders 2 streets on the north and the east, and in the east (a longitudinal border) there is a T-junction and a an open-air parking lot on a culvert. This area has a peculiar atmosphere of openness. Signboards and plant pots are put on the street and there are many long-term residents, who are often seen chatting around the street.
This The site is like a "special valley spot of the zoningregulation". A zoning change line A zone changed-line (Commercial district to Housing district) is designated near the southern border line of the lot.

fig.1

So, the house can be was built slightly higher than the neighboring one in this site. neighbor's. Under such surroundings, a two-story wooden house is built. A two-story wooden house is built under the surroundings.
The 1st floor which connects is connected to the street for creating living space is linked with lively outside, and the 2nd floor is taken a physical distance from the busy street and its floor slab is lifted higher than the neighboring eaves for taking enough sunlight into the 1st floor.
In winter, the kamakura-shaped hill with a bathroom inside can store solar heat collected through a large window allocated upper south part on the 1st floor. It can also be functioned as various living goods such as a sofa when watching TV, a verandah when sitting from outside with a window open, a screen from the entrance, a storehouse and a play place for children. Bedrooms of the 2nd floor are minimum size so the 2nd floor hall is used for all activities such as study studying or reading other rather than sleeping. From the 2nd floor window, an above roof view completely different from the 1st floor opens up (fig.1).
This spatial composition linked up a series of motions in a house with the movement of wind, heat, users and goods may only be is called "two-story wooden house" at present. at the moment. However, I think it is different from the usual one because there is no appropriate name for this composition. at present.

Heating Environment

In winter, sunlight is stored in the concrete hill, collected from the large window allocated under the 2nd floor slab and upper south part on the 1st floor designed for receiving sunlight it and solar heat. The position of the slab and the height of the hill is determined with referencing to the sun's altitude in midwinter and the neighboring eaves height. Floor heating is installed in the concrete hill and the flat area to prepare for days not suitable for heat storage or for the area where the sun's rays do not strike for seasonal conditions.
In summer, sunlight is shaded up by an electrical blind and heated air is stored between windows, which will be is exhausted from the 2nd floor window, through the ventilation slit acquired by installing double wall in the north, the south and roof. In winter, the ventilation slit can be used as the heat storage and supplies warm air through air inlets (fig.2).

Hill Cover

The hill cover is designed as a set of the hill as a method to deal with the regeneration risk of concrete and slow start-up of the floor heating. The cloth commonly used for sofa is sewn in two-ply and urethane is placed in the cushion, which improves heat storage and durability. Rather than using draping, the plane form method is used for making the 3D curved surface of the hill. It is designed by using a drawing and a model modeling and finally measurements are taken by using sample cloth (fig.3). Pockets on the cover can be used for holding mobile phones, remote controls or newspapers and if it is needed more, pockets it can be easily installed more.

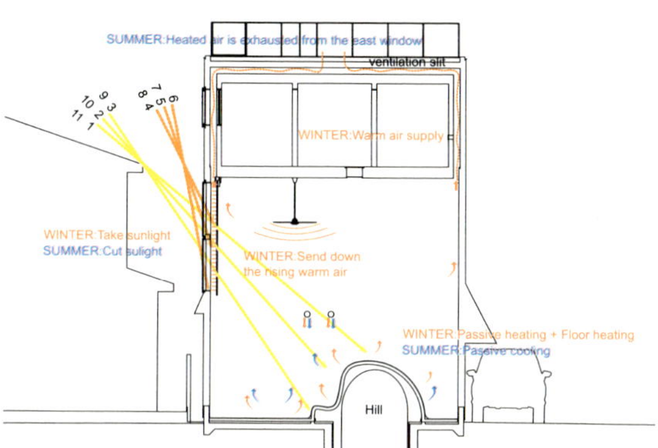

fig.2

fig.3

62-00

View of the south facade. Solar heat is gathering through a large south window.

SPACESPACE

The hill can store solar heat gathered through a large window allocated upper south part on the 1st floor.

The 1st floor connects to the street for creating living space linked with lively outside.

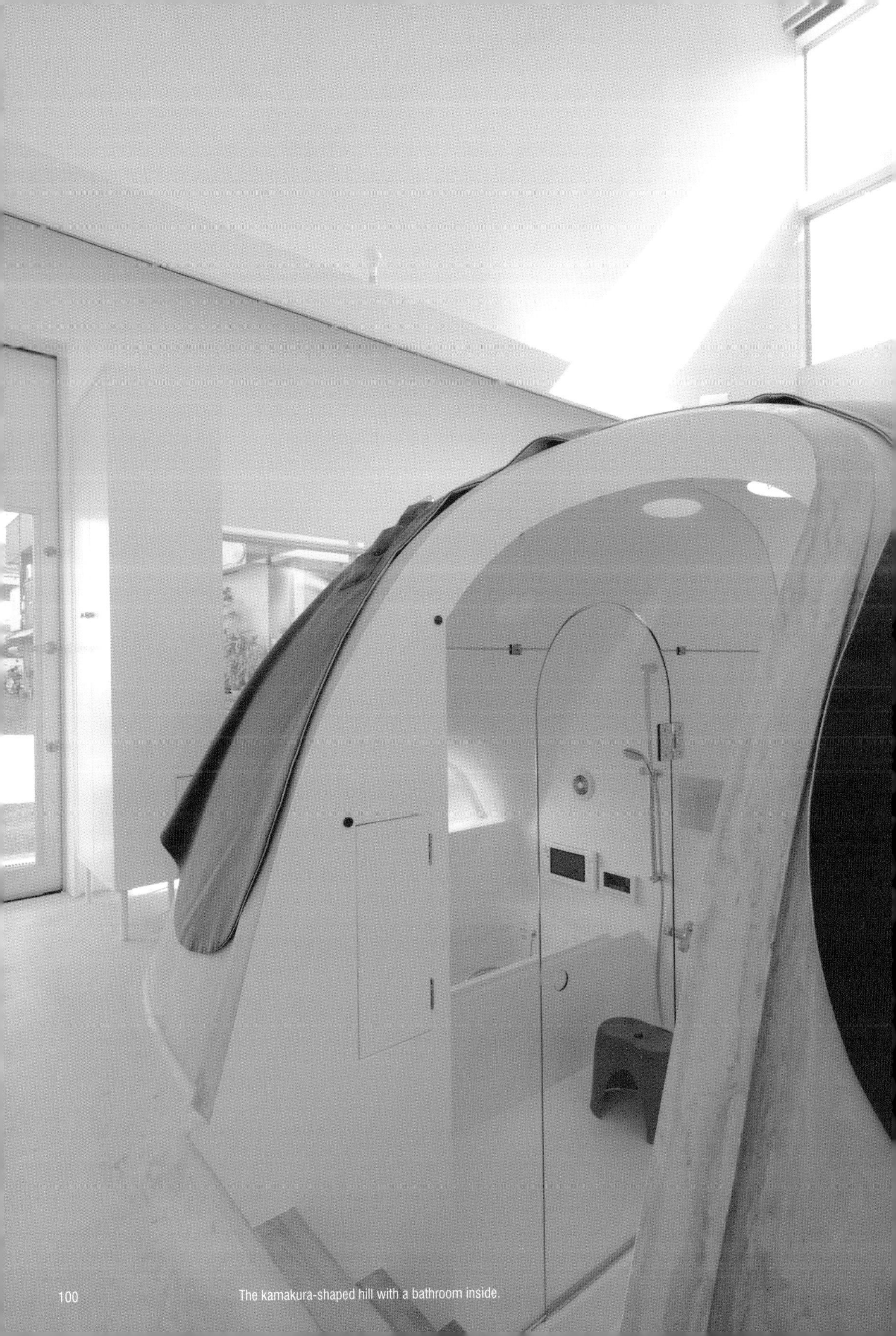

The kamakura-shaped hill with a bathroom inside.

View from bathromm to lavatory in the hill.

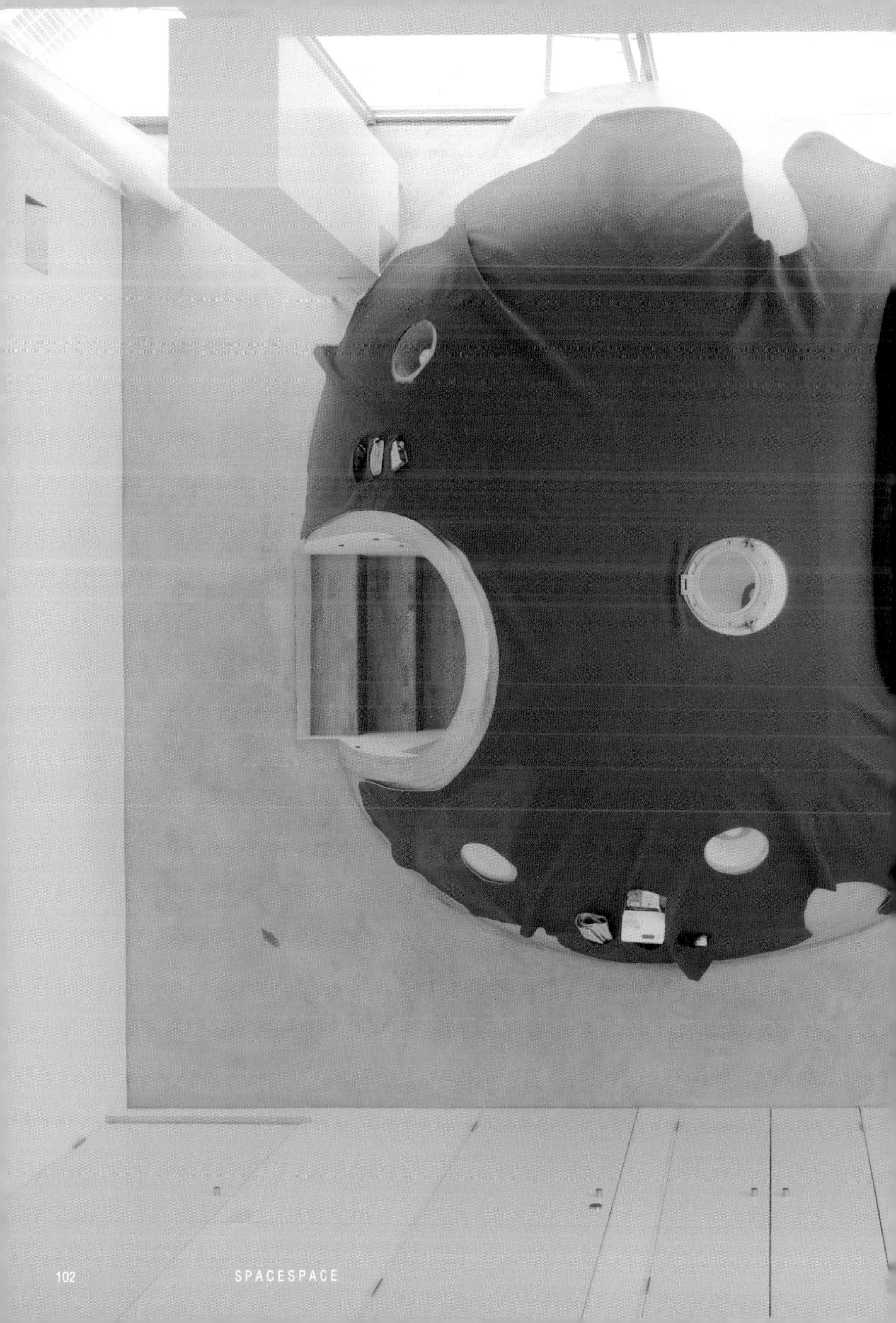

View from hall of the 2nd floor to bedrooms and stairs.

View from stirs to hall.

The above roof view can be seen from the 2nd floor windows.

Downward view of stairs.

upward view of stairs.

A
STR
DK
L
BR
P
E
LV
BR
SEC1
STR
LDK
BR
P
SEC2
B
BR
L
DK
STR
BR
P
SEC1
BR
DK
STR
P
SEC2
C
BR
L
DK
STR
BR
E
LV
F
BR
BR
BR
LFT
LDK
STR
BR
LV
BTH
P
SEC1
BR
BR
LFT
LDK
STR
LV
E
SEC2
G
STR
STR
BR
TR
BR
LDK
LV
P
SEC1
TR
BR
LDK
TR
SEC2
H
LDK
BR
P
BR
K
BR
L winter
K
BR
L summer
LV
P
SEC1
BR
K
BR
LV
SEC2
L
STR
BR
BR
P
LDK
SEC1
STR
BR
BR
LV
LDK
SEC2
M
BR
STR
LD
K
BR
BR
P
P
BR
HAL
BR
LDK
HILL
BTH
LV
SEC1
BR
BRR
LDK
HILL
LV
SEC2
Q
BR
HAL
BR
LDK
HILL
LV
BR
BR
LDK
HILL
LV
BTH
R
BR
HAL
BR
LDK
HILL
LV
U
TR
BR
SR
STR
BR
LDK
HILL
DS
BN
BTH
SEC1
TR
BR
BR
LDK
HILL
DS
LV
BTH
SEC2
V
TR
BR
STR
HALL
SR
BR
LDK
HILL
BN
DS
BTH
SEC1
BR
BR
LDK
HILL
DS
LV
BTH
SEC2
W
TR
BR
HALL
STR
BR
LDK
HILL
BN
DS
BTH

STUDIES 1/500

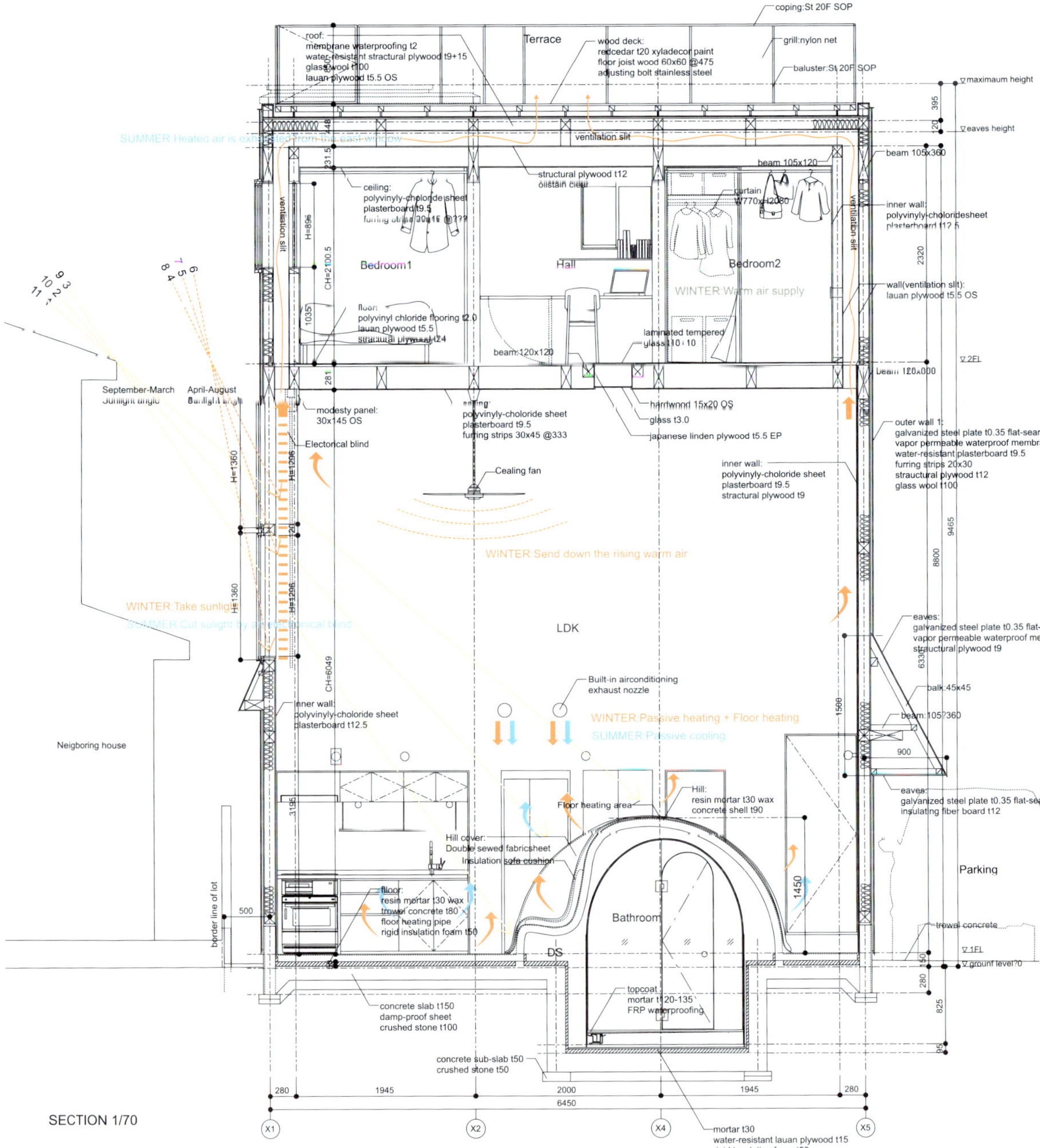

SECTION 1/70

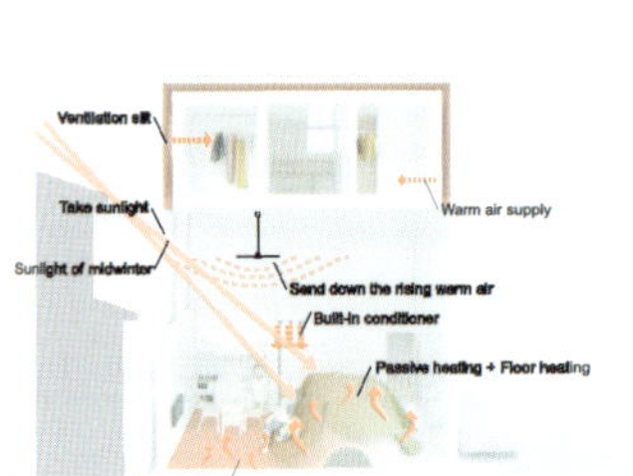

WINTER

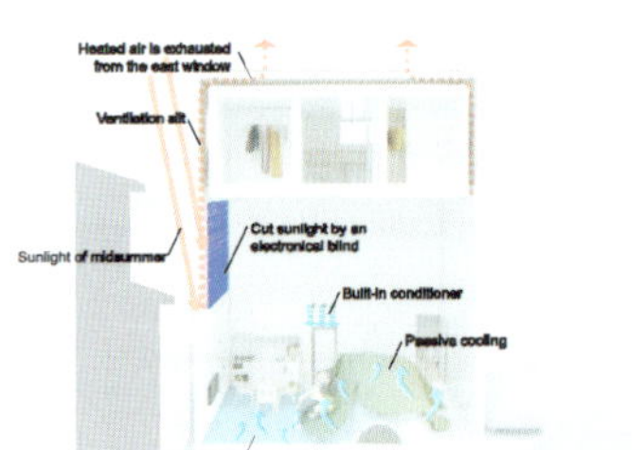

SUMMER

Floor heating and Passive heating by sunlight

WINTER

Cutting sunlight and air axhaust by gravity ventilation

SUMMER

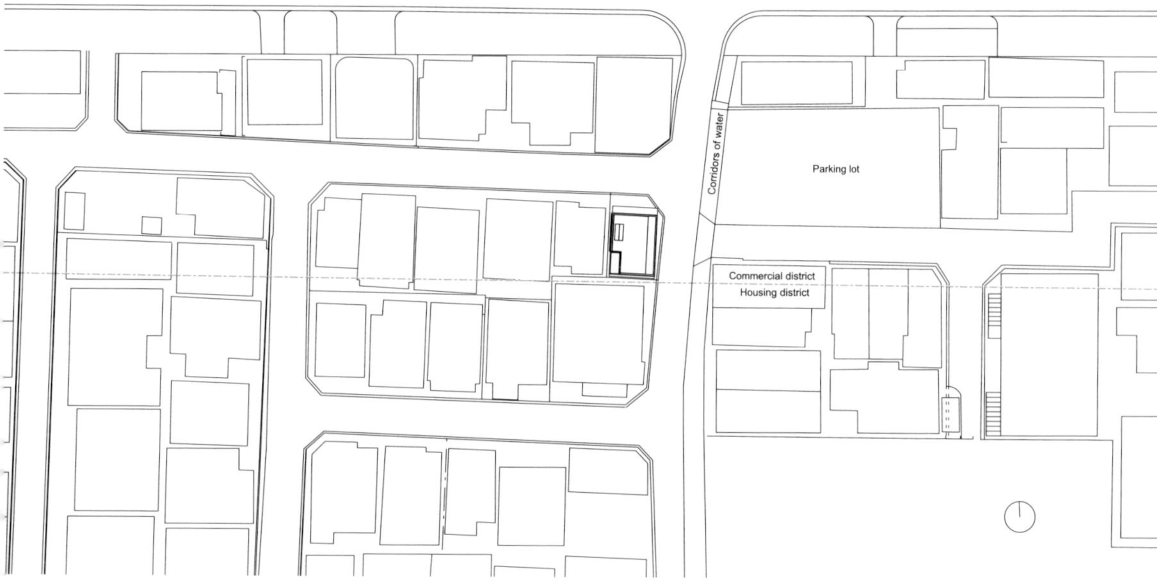

SITEPLAN 1/800

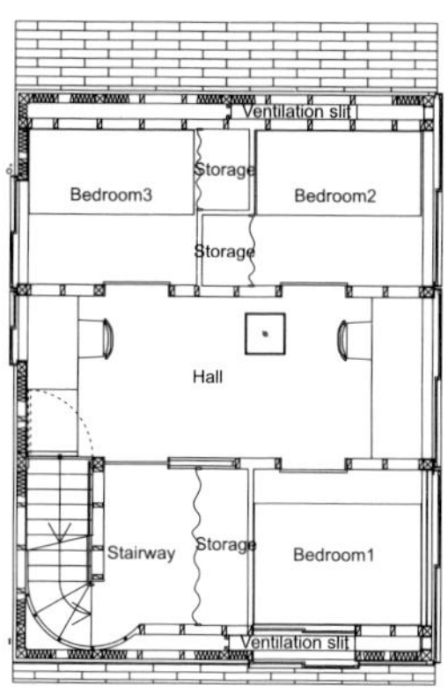

nd FLOOR PLAN 1/150

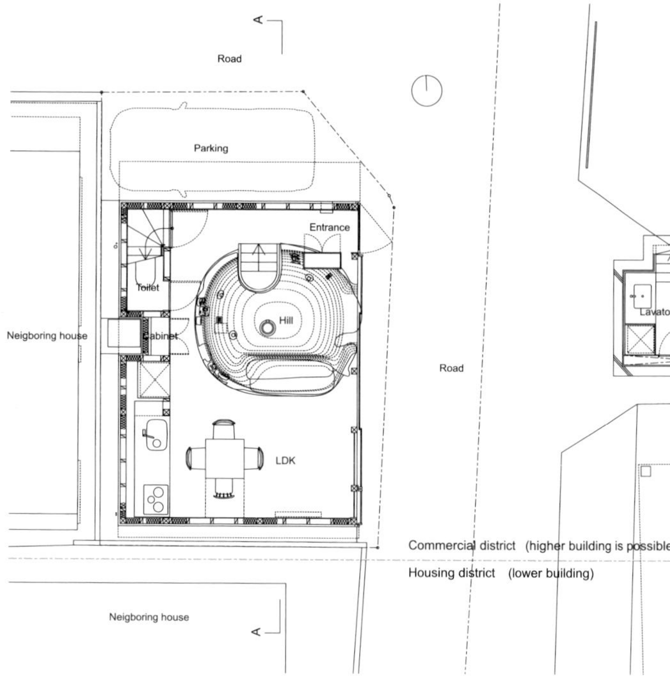

1st FLOOR PLAN 1/150

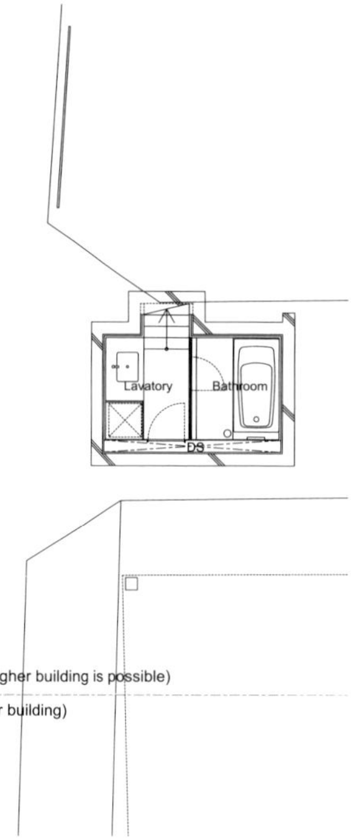

BATH FLOOR PLAN 1/150

AST ELEVATION 1/300

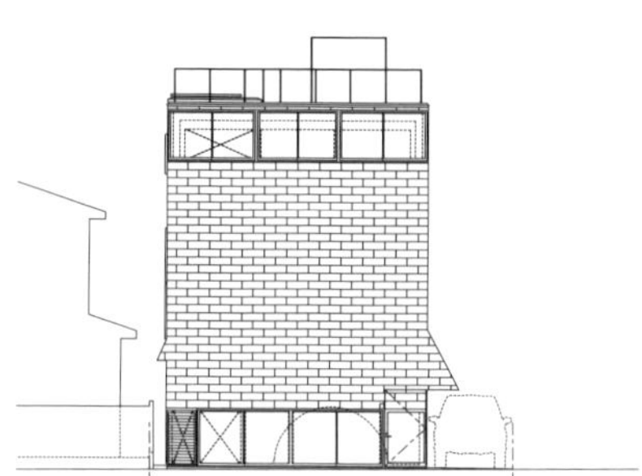

SOUTH ELEVATION 1/300

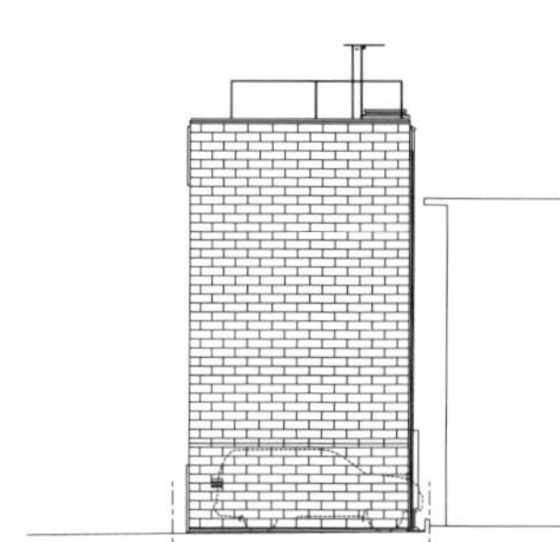

NORTH ELEVATION 1/300

Ground and above roof house

location: Ibaraki-city, Osaka
principal use: Residence
family composition: 2+2

design

design: SPACESPACE
Takanori Kagawa, Junko Kishigami
Yu Goto(enviromental advice)
stractural design: TAPS
Masaichi Taguchi, Shinsaku Kataoka
general constractors
Kohatsu Co.,Ltd. Takahiro Kinugawa
curtain+hill cover
fabricscape, Kiyohiko Yamamoto
truss-wall
asashi building-wall
Photo: Koichi Torimura

area

site area: 53.31m²
building area: 30.96m²
total floor area: 61.92m² all(73.37m²)
1stfloor 35.96m² 2ndfloor 30.96m²
building-to-land ratio: 58.08% (90%)
floor-area ratio: 125.52% (240%)
number of stories: +2

dimensions

maximum height: 9,409mm
upper eaves edge height: 8,894mm
story height: 1st(hall): 6,330mm
2nd(hall•bedroom): 2,320mm
ceiling height: 1st(hall): 6,049mm
2nd(hall•bedroom): 2,100mm

site conditions

district: neighborhood commercial district
road width: east5.62m, north4.0m
parking lot: 1

structure

structure: wood
basis: independent footing

equipment

air-conditioning:
all-in-one air-conditioning system,
built-in air conditioner, floor heating
thermal source: electricity, sunlight

process of works

design period: 2009/8-2010/3
construction period: 2010/4-2010/9

outside finishing

roof: wood deck,redcedar t20 xyladecor pain
+membrane waterproofing t2
outer wall: galvanized steel plate t0.35
window: aluminum sash
exterior: trowel concrete

inside finishing

[1Fhall]
floor: resin mortar t30 wax
wall: polyvinyly-choloride sheet
+plasterboard t12.5
ceiling: polyvinyly-choloride sheet
+plasterboard t9.5
[toilet]
floor: resin mortar t30 wax
wall: polyvinyly-choloride sheet
+plasterboard t12.5
ceiling: polyvinyly-choloride sheet
+plasterboard t9.5
[bath room+lavatory]
floor: FRP waterproofing
wall: insulating fiber board t6+6 VP painting
ceiling: mortar t20 VP paiting
[2Fhall+bedroom]
floor: polyvinyl chloride flooring t2.0
wall: polyvinyly-choloride sheet
+plasterboard t12.5
ceiling: polyvinyly-choloride sheet
+plasterboard t9.5

used equipment

air conditioning: Toshiba, Panasonic
sanitary: TOTO INAX Panasonic ADVAN
kitchen: Hitachi, Betty&Home
lighting: Saita, Panasonic

Night view.

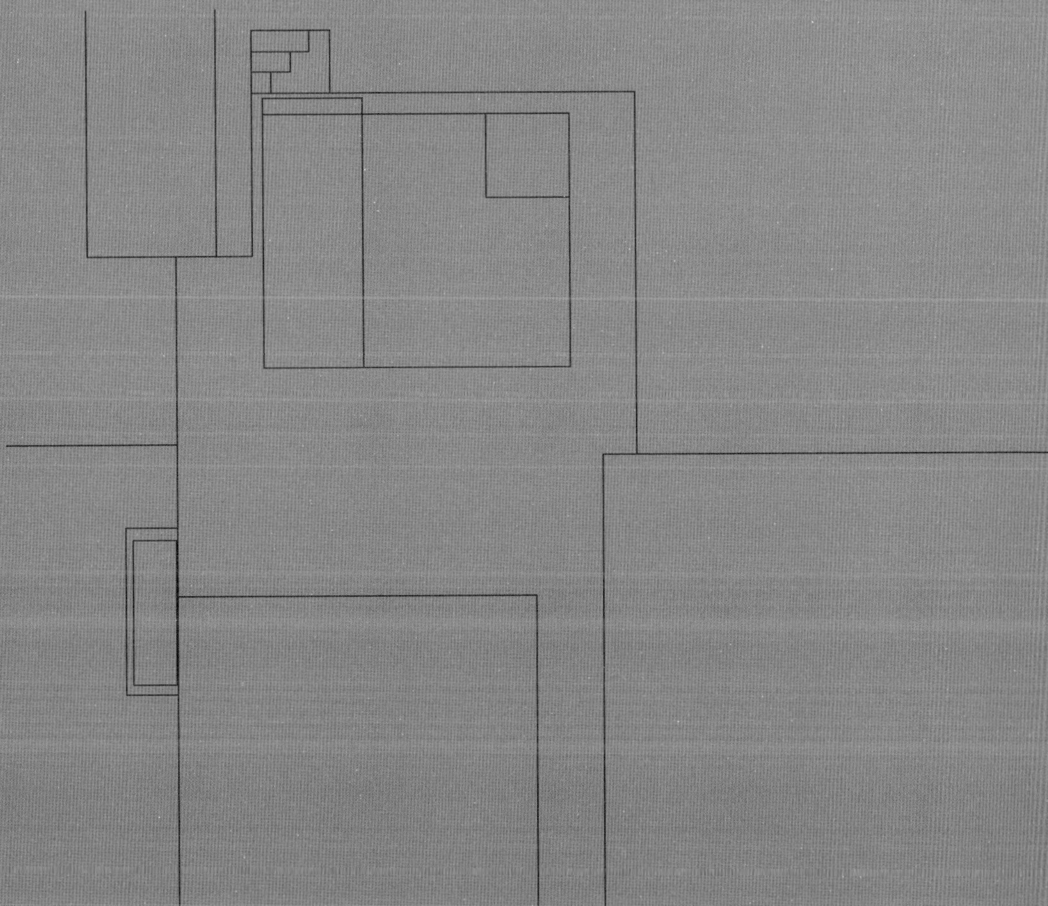

LONGTALL HOUSE

0611

location:	Meguro-ku, Tokyo
principal use:	residence
site area:	66.49m^2
building area:	39.58m^2
total floor area:	102.69m^2
structure:	W+RC
number of stories:	-2+3
completion:	November, 2006

길고 키 큰 집

위치는 전형적인 일본의 주거 밀집지역으로서 시부야에 인접한 두 개의 역 사이에 있다. 결혼한 젊은 부부를 위한 이 주택은 분할 거래된 띠 형상(약 4m x 16m)의 대지를 가진다. 대지는 남측과 북측에 있는 길들과 접하면서 그 둘을 연결시킨다. 북측 길을 따라 연속되어 있는 낡은 옹벽은 3.5m 높이의 절벽을 이루고 있다. 이 길은 현재 기준으로는 너무 좁기 때문에, 좀 더 넓히기 위해서는 일부 벽의 제거가 요구되었다. 이 낡은 벽은 부분적으로 새로운 구조로 교체되었고, 우리는 이를 통해 주택을 구성해보고자 하였다. 실제의 상황에 적용된 구조적 변형은 벽체보다도 더 오래된 지형적 조건을 고려하는 것으로서 두 개의 지하층이 만들어졌다. 이 주택은 두 개의 높이가 다른 지면을 연결시키면서, 주변 환경과의 연속적 관계성을 높일 수 있었다.

이 주택은 모두 다섯 개의 층을 갖는다. 아래의 두 층은 토압을 버티기 위한 철근콘크리트로 되어 있고, 위의 세 층은 경량화를 위한 목구조로 되어있다. 1층은 침실, 욕실, 주차장과 기타공간들로 밀도 있게 구성되어 있고, 2층은 테라스와 거실, 그리고 주방 식당을 가지면서 좁은(그러나, 통풍이 잘되는) 공간으로서, 독특한 모양의 작은 땅에 대응하고 있다.

이 주택은 두 개의 건축을 포함하고 있다. 하나는 "long row house(길게 연속된 주택)"와 같은 "긴 주택(long-house)"이고, 다른 하나는 북측 벽을 따라 연속된 타워 같은 "키 큰 주택(tall-house)"이다.

LongTallHouse

The site is located in typical Japanese crowded crowded Japanese's residential area, just middle between two railway stations near to Shibuya.

The house of a young married couple is on a strip of site subdivided for sale (about 4mX16m). It adjoins two paths at north and south sides, and locates on a very old retaining wall which runs along the north path and forms 3.5m height cliff. The path was too narrow to comply with the current regulations, so we needed removing to remove a part of the wall so that it could be widened. make widely.

The old wall was partially to be replaced with a new structure, so we tried to merge the house with it. Our structural adjustment to actual condition brought 2 basement stories which trace old topography older than the wall. The house got a link between two different ground levels and a profound continuity of it and surroundings.

The house has 5 stories. The lower 2 floors are made of RC to resist soil pressure. The upper 3 floors are made of wood timber structure for lightness. The 1st floor has bedroom, bathroom, parking and others compactly. The 2nd floor has terrace, living room and , dining kitchen and in a narrow (but airy) space responded distinctive shape of the plot.

This house includes 2 buildings; one is a long-house like a long row house, the other is a tall-house like a tower continuing to thc north rctaining wall.

Longhouse in Vietnam

Tallhouse

Longhouse in Izumosaki, Japan

Tallhouse inTokyo, Japan

Longhouse in Sumiyoshi, Japan

Tallhouse inTokyo, Japan

Longhouse in Kyoto, Japan

Tower maison inTokyo, Japan

Repark
月極駐車場

North facade.

Moving gravity type fittings changes the facade.

South facade.

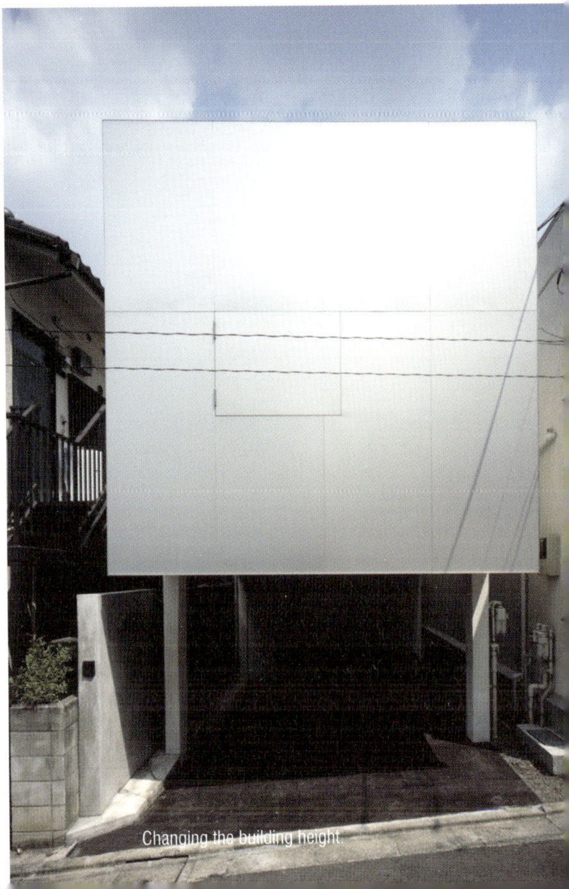

Changing the building height.

Circular staircase passes through 5 layers (B2F to 3F).

View of living room.

View from B2F to the back side of north facade.

B2F guestroom.

View from lavatory to bathroom.

1st floor corridor.

Aerial view.

STUDIES 1/600

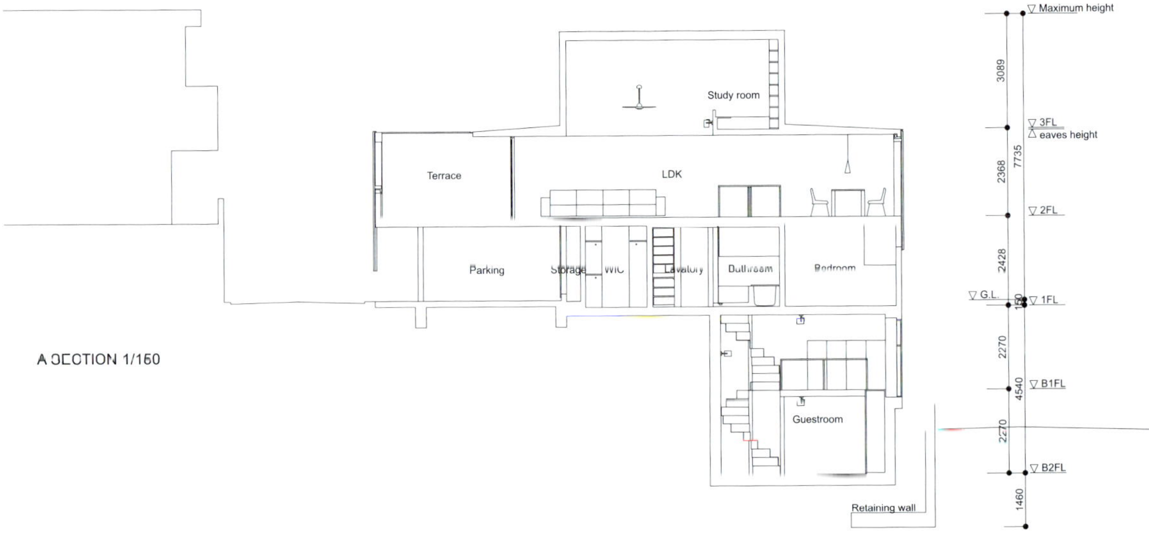

A SECTION 1/150

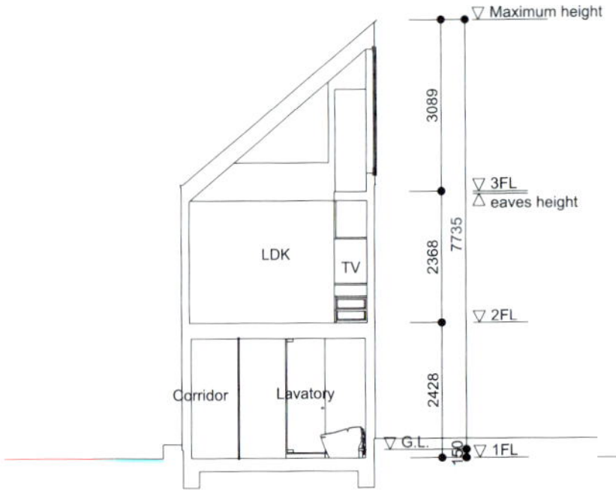

B SECTION 1/150

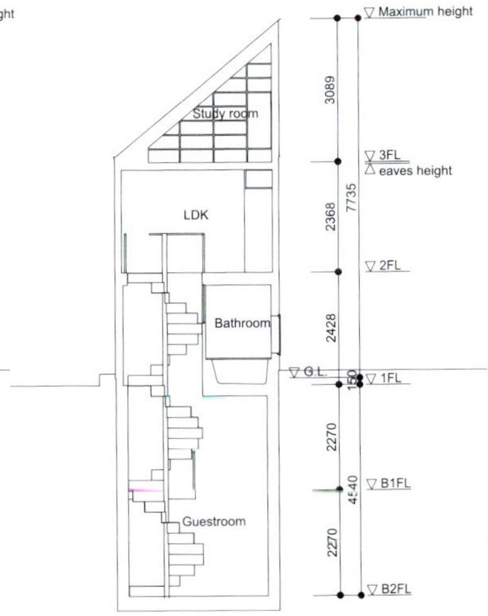

C SECTION 1/150

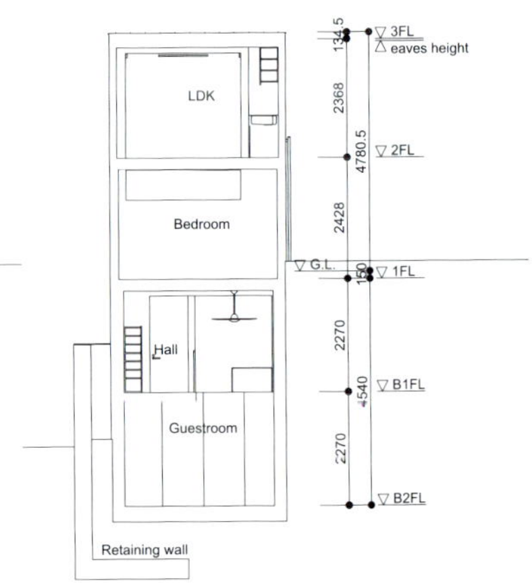

D SECTION 1/150

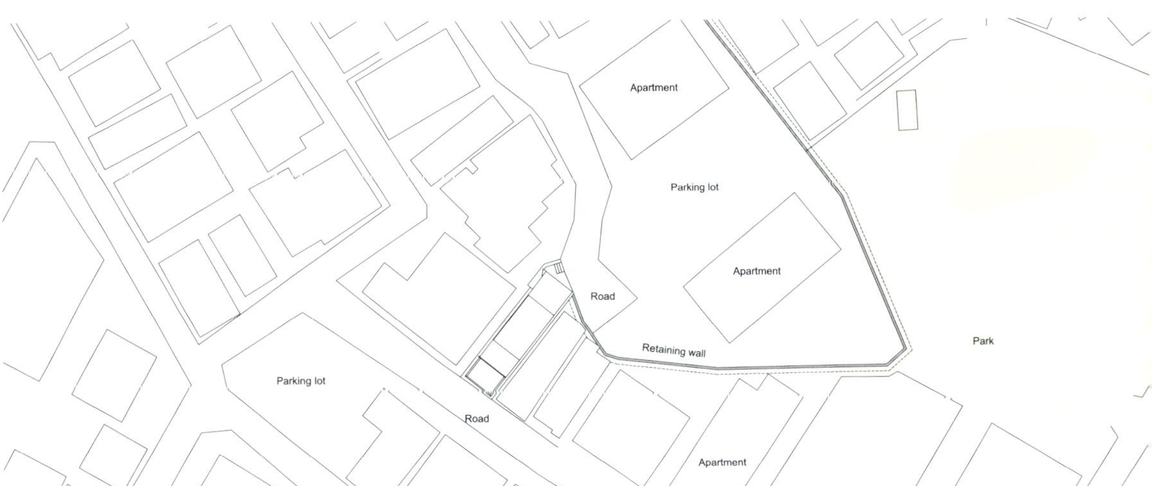

SITEPLAN 1/800

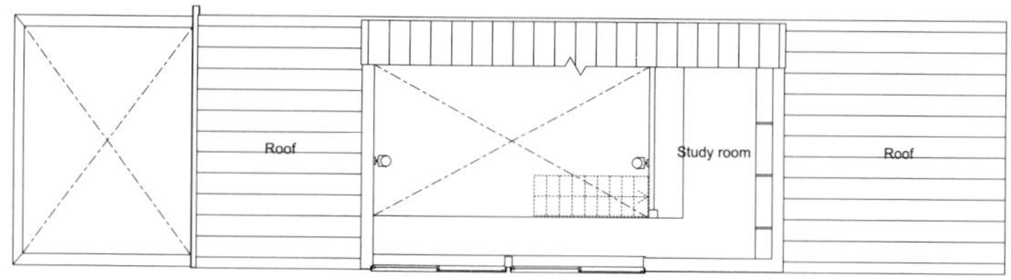

3rd FLOOR PLAN 1/150

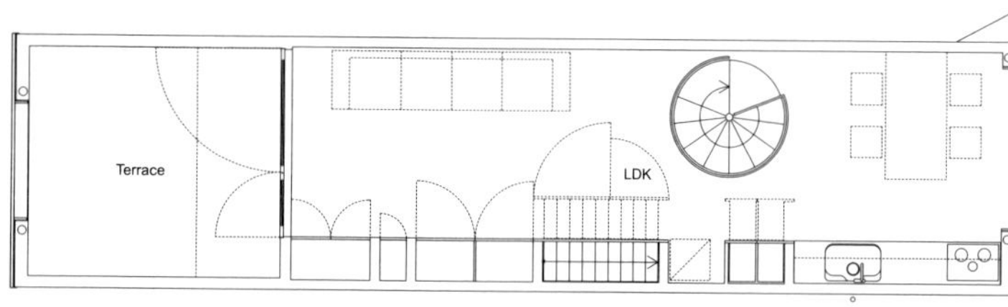

2nd FLOOR PLAN 1/150

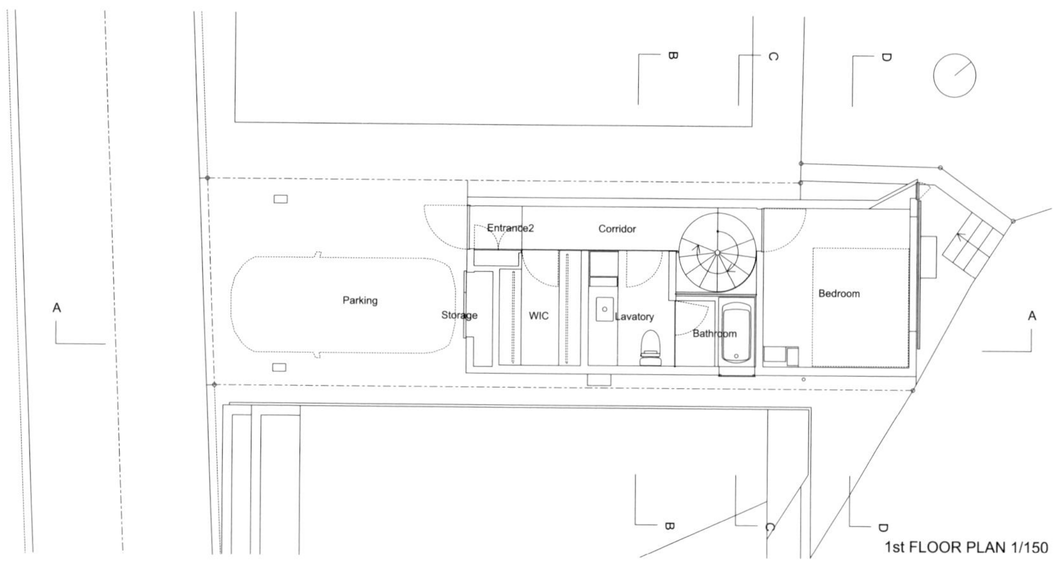

1st FLOOR PLAN 1/150

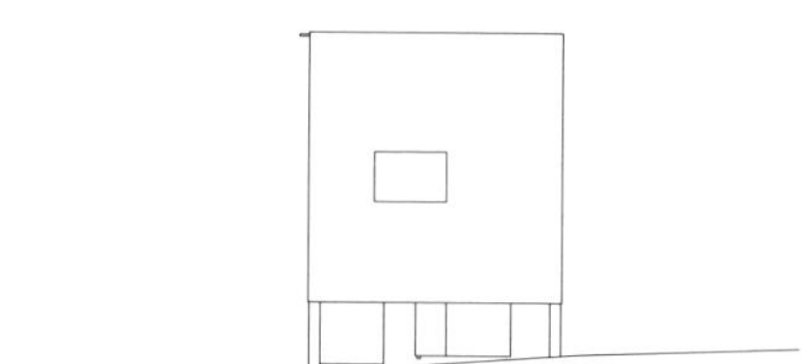

SOUTH ELEVATION 1/200

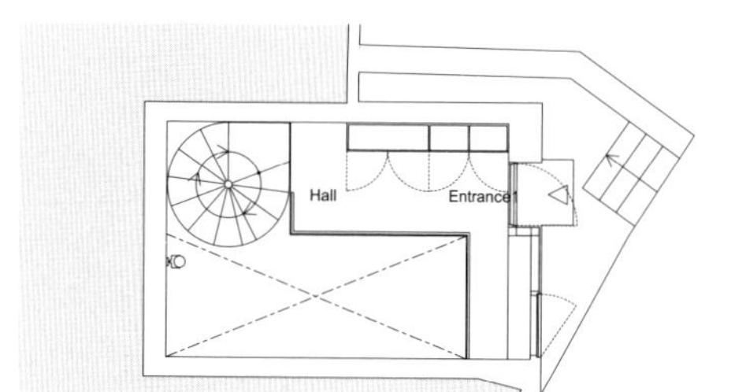

B1st FLOOR PLAN 1/150

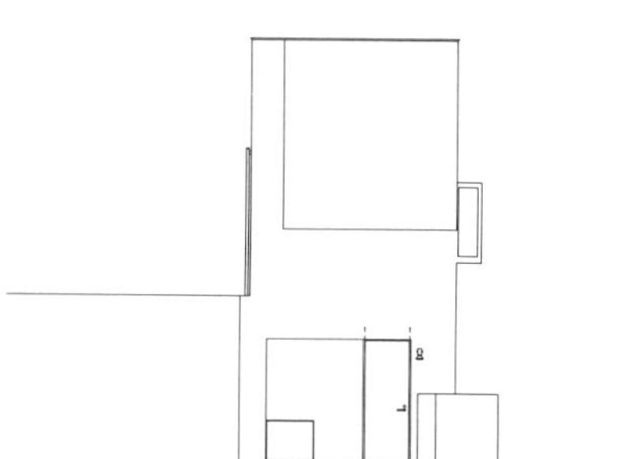

NORTH ELEVATION 1/200

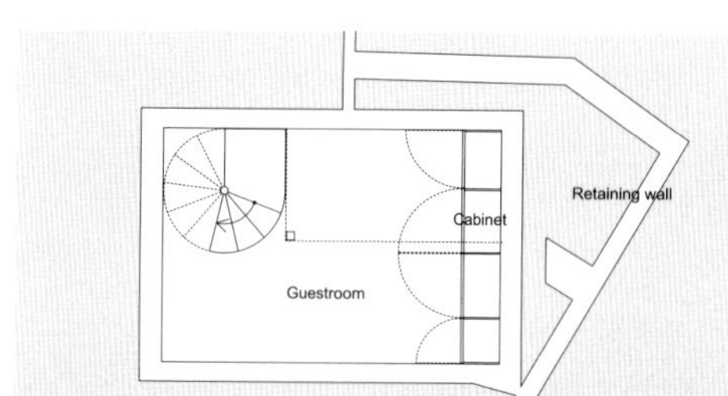

B2nd FLOOR PLAN 1/150

月極駐車場

Longtallhouse

location: Meguro-ku, Tokyo
principal use: Residence
family composition: 2

design
design: SPACESPACE
Takanori Kagawa
stractural design:
Nobutaka Kashimoto
general constractors
Nakano Co.,Ltd. Yukio Yoshimura
doors and windows
Inoue kouki Co.,Ltd. Eiji Inoue
Photo: Koichi Torimura

area
site area: 66.49m²
building area: 39.58m²
total floor area: 77.76m²
3rdfloor 9.06m²
1stfloor 32.44m² 2ndfloor 36.26m²
B1stfloor 9.04m² B2ndfloor 16.60m²
building-to-land ratio: 59.53% (60%)
floor-area ratio: 116.95% (150%)
number of stories: -2+3

dimensions
maximum height: 7,735mm
upper eaves edge height: 4,600mm
story height: 3rd: 3,089mm 2nd: 2,368mm
1st: 2,428mm B1st: 2,270mm B2nd: 2,270mm
ceiling height: 2nd: 2,200mm
1st: 2,200-2,100mm B1st: 2,040mm
B2nd: 4,310-2,094mm

site conditions
district: low building residential district
category 1 height district
road width: south4.0m, north4.0m
parking lot: 1

structure
structure: RC + wood
basis: mat foundation

equipment
air-conditioning:
all-in-one air-conditioning system
thermal source: gas

process of works
design period: 2005/11-2006/4
construction period: 2006/5-2006/12

outside finishing
roof: galvanized steel plate t0.35 standing -seam roofing
outer wall: mortar t30 + lithing spraying
window: aluminum,wood,steel sash
exterior: trowel mortar, gravel

inside finishing
[entrance1]
floor: polyvinyl chloride flooring t2.0
wall: plasterboard t12.5 EP painting
ceiling: exposed concrete EP painting
[entrance2 + corridor]
floor: resin mortar t30 wax
wall: plasterboard t12.5 EP painting
ceiling: plasterboard t9.5 EP painting
[LDK]
floor: iron-wood flooring t15 wax
wall: plasterboard t12.5 EP painting
ceiling: plasterboard t9.5 EP painting
[bedroom]
floor: pine flooring t15 wax
wall: PBt12.5 + polyvinyly-choloride sheet
ceiling: PBt9.5 + polyvinyly-choloride sheet
[guestroom]
floor: resin mortar wax
wall: exposed concrete EP painting
ceiling: plasterboard t9.5 EP painting
[lavatory,toilet,bathroom]
floor: porcelain tile 150x150 t6
wall: porcelain tile 150x150 t6
insulating fiber board t12 VP painting
ceiling: insulating fiber board t12 VP painting
[WIC]
floor: resin mortar wax
wall: PBt12.5 + polyvinyly-choloride sheet
ceiling: PBt9.5 + polyvinyly-choloride sheet
[storage]
floor: pine flooring t15 wax
wall: plasterboard t12.5 EP painting
ceiling: plasterboard t9.5 EP painting

used equipment
air conditioning: DAIKIN
sanitary: TOTO CERA GRHOE Tform
kitchen: Herman Chugaikoueki
lighting: Yamagiwa Yamada Endo Panasonic

Night view.

ISOYAMA DISPENSING PHARMACY

1010

location: Suzuka-city, Mie

principal use: pharmacy

site area: $168.77m^2$

building area: $77.88m^2$

total floor area: $65.75m^2$

structure: S

number of stories: +1

completion: October, 2010

이소야마 조제약국

이 프로젝트는 일본 미에현 스즈카 노변에 위치한 조제약국이다. 스즈카는 기차로 오사카에서 2시간, 나고야에서 한 시간 정도 떨어진 일본의 전형적인 지방 도시이며, 스즈카서킷으로 유명하다.

자동차가 일상생활에서 일반적으로 사용되는 일본의 지방 도시에서는 상대적으로 상업시설에 큰 주차시설이 필요하다. 따라서 많은 상업시설은 기차역 근처보다는 고속국도를 따라 위치하는 경향이 있다.

이런 환경에서 건물의 전면이 도로에 연속해서 셋백되어 있고 차들의 행렬이 길을 따라 끊임없이 놓여있는 전형적인 도로 풍경을 종종 볼 수 있다. 이런 관습적인 모습에서는 도시만의 독특한 풍경을 만들기 어렵다.

만약 작은 상점이 주거지역에 위치한다면 입면의 연속성을 파괴한다. 그러나 그렇다고 그 작은 크기나 사용의 편리성을 변경하긴 어렵다. 그래서 조제약국은 건물들보다는 주차구획으로 구성된 전형적인 노변풍경을 변화시키고자 하는 방법으로 설계되었다.

조제약국 건물은 다른 건물들과 마찬가지로 주차구역 뒤에 위치한다. 그러나 주차구획은 사용자와 지역주민인 열린 광장으로 사용할 수 있도록 설계되었다.

꽃무늬 패턴은 교통표시 페인팅과 함께 콘크리트 위에 새겨진다(그림 1). 외부공간에서 패턴들은 점으로 이루어진 선들로 차량흐름을 조절하고(주차구획, 보행지로, 서비스존), 사용자와 지역주민을 위한 모임의 장소, 이이들의 놀이 장소를 제공한다.

더욱이 꽃무늬 패턴은 카스토퍼와 기둥으로 압출성형된다. 이 패턴들은 (장식적인 디테일에서 고대 기둥들과 달리) 구조적으로 계산된 수치적인 형태로 변환된다(그림 2).

어떤 기둥은 3.2밀리미터 강철판을 구부리고 용접하여 만들어졌다(그림 3). 이 기둥들은 2미터 간격으로 지붕프레임을 지지하고 있다(그림 4). 평면적인 장식은 수직적인 리브로 형상화 된다(그림 5).

주차장은 건축설계를 할 때 일반적으로 우선순위에서 뒤에 있는 부분이지만 이 프로젝트에서는 지배적인 요소로 다루어졌으며, 다시 쓸 수 있는 교통표시 페인트로 디자인되었다. 그래서, 이 프로젝트의 주요부분은 컴퓨터의 어플이 계속해서 업데이트 되듯이, 계속해서 재-디자인될 수 있다.

fig.2

fig.3

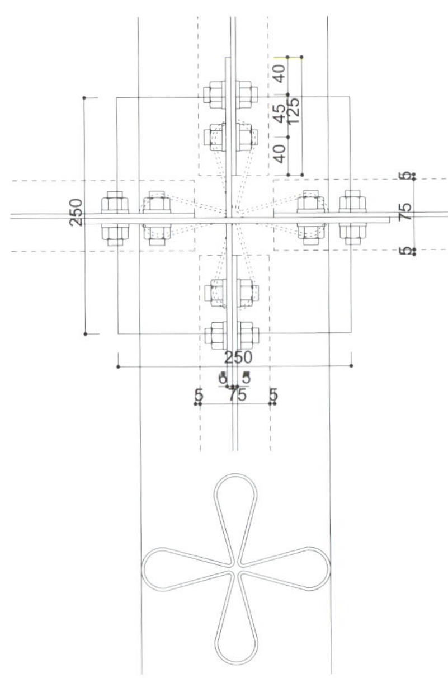

fig.4

Isoyama dispensing pharmacy

This is a dispensing pharmacy situated on a roadside in Suzuka, Mie prefecture, Japan. Suzuka is a typical local suburban city in Japan about two hour from Osaka, and about one hour from Nagoya by train.

Suzuka is famous for Suzuka Circuit. In Japanese local suburban city where cars are popular in daily use, relatively large parking lots are necessary for commercial facilities.

And many commercial facilities tend to locate along national highways rather than in front of main railway stations.

In these surroundings, the typical conventional road sight, building's facades set back from the road in a row and the line of cars endlessly sets along the road, can be frequently seen. This convention can't make the scenery or a specialty of the particular city.

If a small shop is located in a residential zone, the continuity of facades has broken off. But this is difficult to change for its smallness and the usability.

So, the dispensing pharmacy is designed as a method for changing the typical conventional road sight consisted of parking lots rather than buildings.

The building of the dispensing pharmacy is located behind the parking lot as usual. But the parking lot is designed to be used as an open square for users and local residents.

Flower patterns are described on the concrete ground with traffic paint (fig.1). In the exterior space, described patterns manage the traffic flow as dotted lines (dividing parking positions, walking paths and service zone), and form gathering spots for users and local residents and play spots for children. In the interior space, these patterns also control the users flow.

Furthermore, the flower patterns are extruded as car stops and columns. These patterns (in even their ornamental details, unlike the ancient columns) can be converted into the numerical figures of structural calculation (fig.2).

Some of the columns are made by bending 3.2-millimeter steel plates and welding them (fig.3). These columns support the roof frame at 2-meter intervals (fig.4). The types of horizontal ornaments appear as the number of vertical ribs (fig.5).

Parking lot is a low priority part of designing architecture, but it is dealt as the dominant element in this project, and designed with the rewritable traffic paint. So, the dominant part of this project can be re-designed over and over again, like pc-applications are updated constantly.

fig.1

fig.5

磯山調剤薬局
処方せん受付
TEL 059-380-1100

SPACESPACE

カップルーム

View from front counter to reception room and parking lot for visitors.

View of piloti and reception room.

View of parking lot.

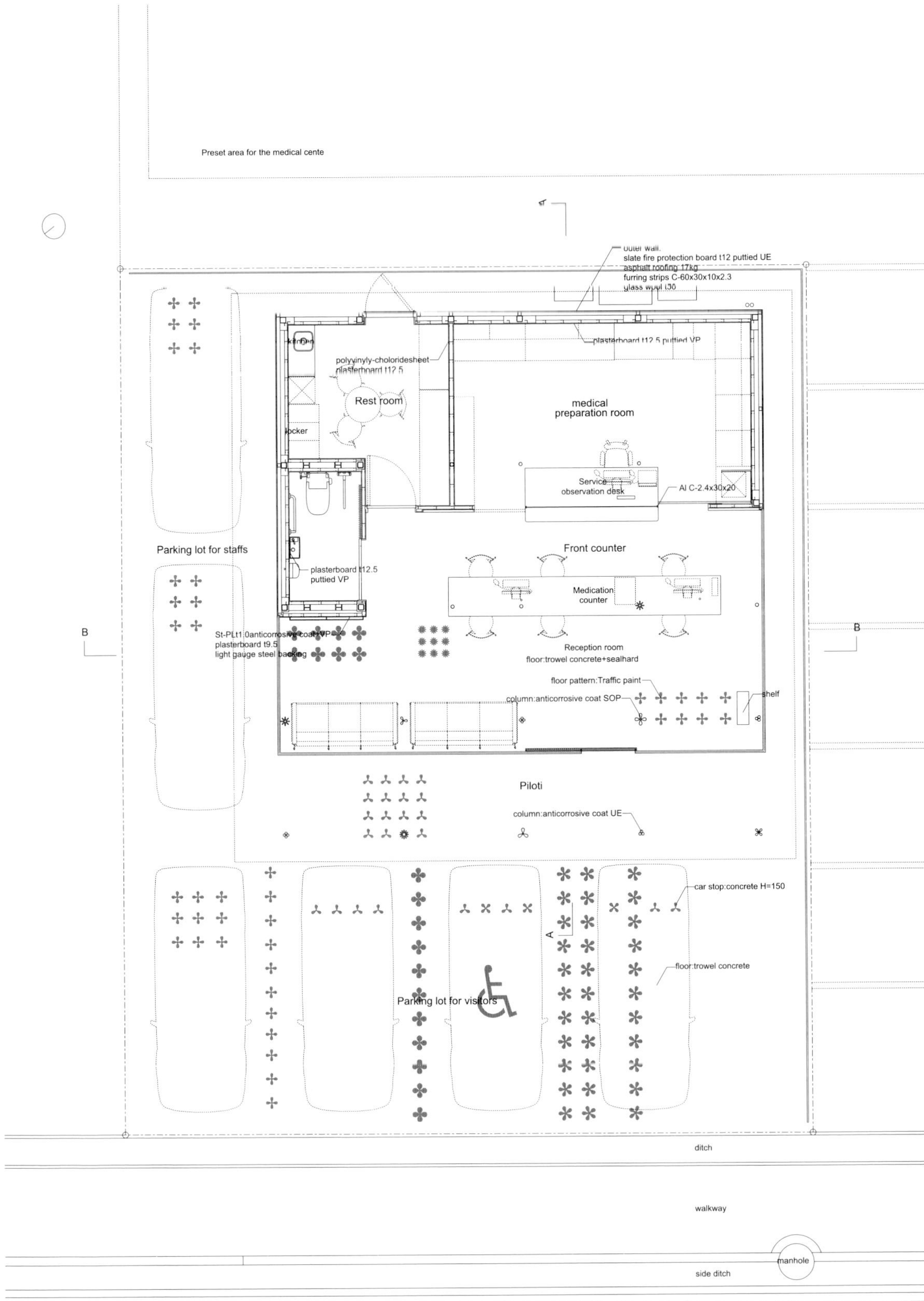

1st FLOOR PLAN 1/00

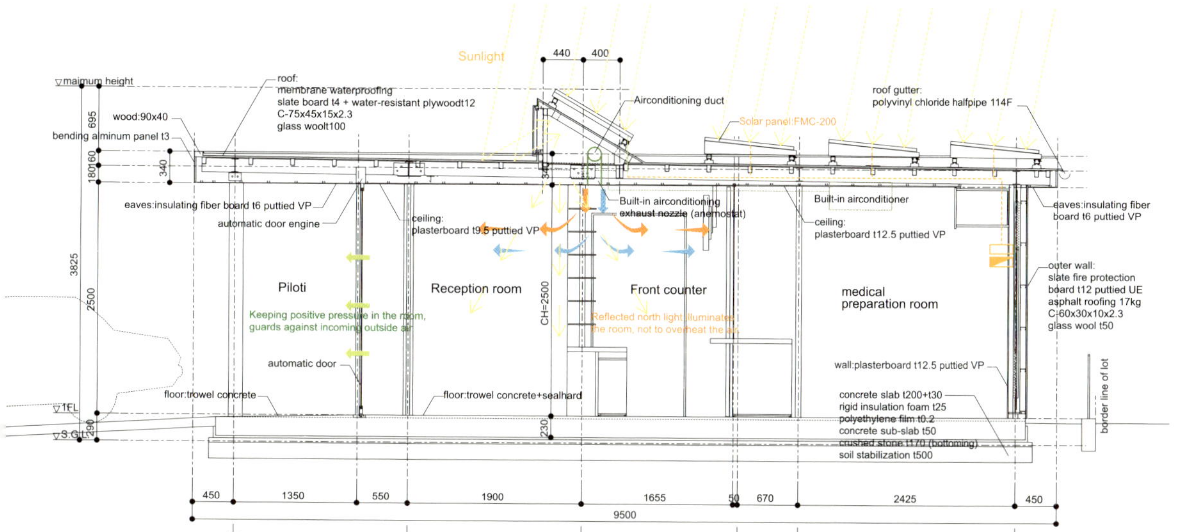

B SECTION 1/80

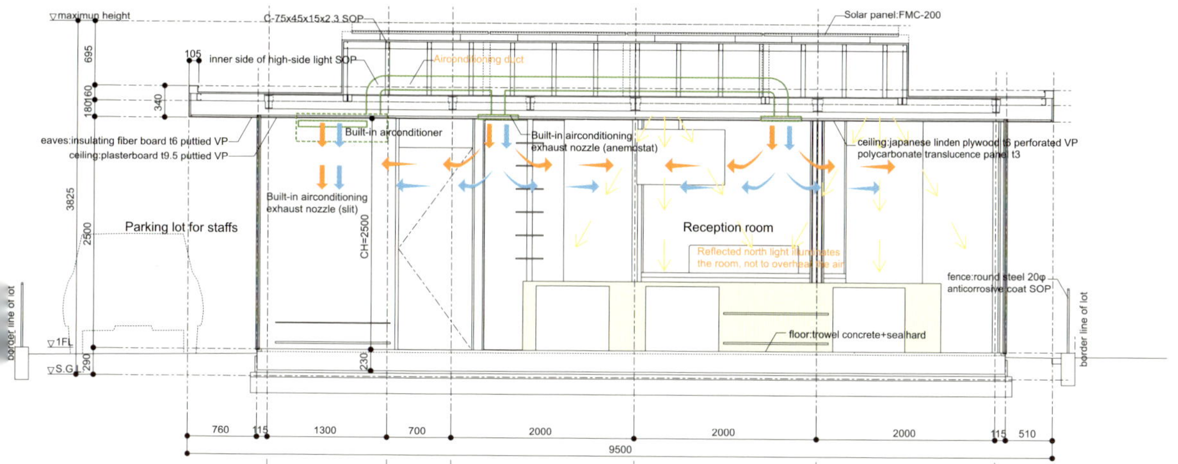

A SECTION 1/80

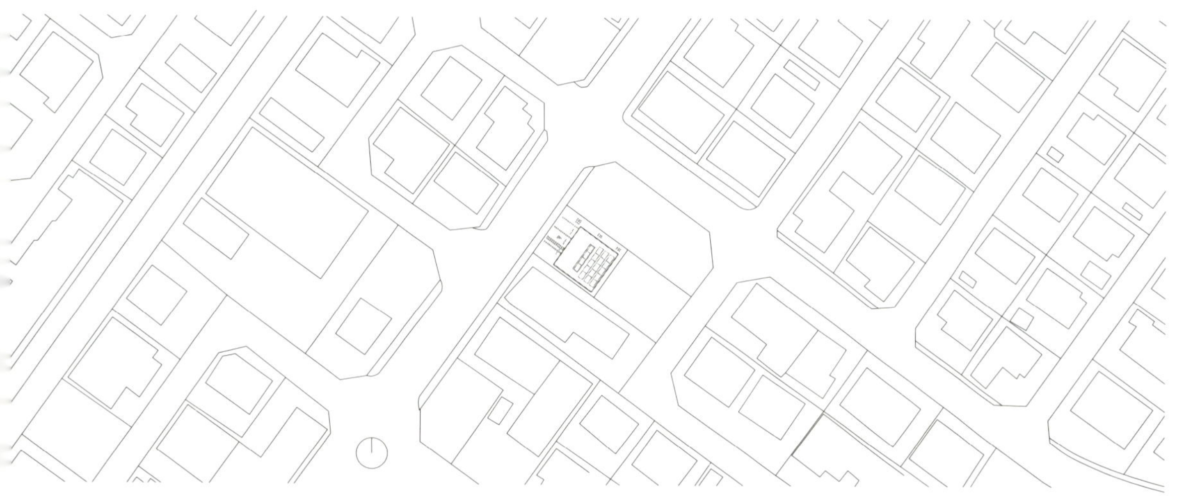

SITEPLAN 1/1500

Isoyama dispensing pharmacy

location: Suzuka-city, Mie
principal use: Dispensing pharmacy

design
design: SPACESPACE
Takanori Kagawa, Junko Kishigami
Yu Goto(enviromental advice)
structural design: TAPS
Masaichi Taguchi
general constractors
Ikeda act Co.,Ltd
flower shape columns
SunWind Co.,Ltd., Yukio Ishida
Photo: Koichi Torimura

area
site area: 168.77m²
building area: 77.886m²
total floor area: 65.75m²
1stfloor 65.77m²
building-to-land ratio: 46.14% (80%)
floor area ratio: 38.96% (200%)
number of stories: +1

dimensions
maximum height: 3,225mm
upper eaves edge height: 3,056mm
story height: 1st: 3,056mm
ceiling height: 1st(reception): 2,500mm
1st(medical preparation room): 2,500mm
1st(rest room): 2,200mm

site conditions
district: neighborhood commercial district
road width: west7.0m
parking lot: 6

structure
structure: steel
basis: mat foundation

equipment
air-conditioning:
built-in air conditioner
thermal source: electricity, sunlight

process of works
design period: 2010/1-2010/6
construction period: 2010/7-2010/10

outside finishing
roof: membrane waterproofing t2
outer wall: siding board t12 UE painting
window: aluminum,steel sash
exterior: trowel concrete + seal hard

inside finishing
[reception room]
floor: trowel concrete + seal hard
wall: plasterboard t12.5 VP painting
ceiling: plasterboard t9.5 VP painting
[medical preparation room]
floor: trowel concrete + seal hard
wall: plasterboard t12.5 VP painting
ceiling: plasterboard t9.5 VP painting
[rest room]
floor: trowel concrete + seal hard
wall: polyvinyly-choloride sheet
+plasterboard t12.5
ceiling: polyvinyly-choloride sheet
+plasterboard t9.5
[toilet]
floor: trowel concrete + seal hard
wall: plasterboard t12.5 VP painting
ceiling: plasterboard t9.5 VP painting

used equipment
air conditioning: Mitsubishi DAIKIN
sanitary: TOTO Panasonic
kitchen: Chugaikoueki
lighting: Panasonic

Night view.

HUMANSCAPE

0805

location: Kumamoto-city, Kumamoto

principal use: station square

site area: 5,700m^2

building area: 5,700m^2

total floor area: 5,700m^2

structure: S+Trafficpaint

number of stories: +1

completion: May, 2008

휴먼스케이프

이 프로젝트는 쿠마모토 역 서측 출구 광장 현상설계를 위한 제안이다.
동측 출구 광장(하루에 5만4천명의 사용자)은 슈퍼블록에 놓여있는 거대한 스케일의 상업시설에 둘러싸여 있다.
반대로 서측 광장(하루에 1만8천명의 사용자)은 소규모 상점과 기존 주거단지로 구성된 작은 스케일의 시설과 만나고 있다. 그래서, 서측 출구 광장은 지역주민들에 의해 자주 사용된다. 더욱이 기존 철로가 이 기차역의 동측 부분에 위치하고 있고, 신칸센은 서측부분과 대면하고 있다. 이것은 다른 현에서 오는 사람들은 서측 출구에서 보는 광경이 도시의 처음 보는 모습이라는 것을 의미한다.
우리는 작은 스케일의 시설과 기차역 건물을 사람들이 행위와 통합하는 이음매 없는 풍경을 제안한다. 식물과 같은 형태의 캐노피, 작은 구조물, 바닥패턴 그리고 사용자들의 중첩이 도시풍경을 연속시킨다. 다양한 기능(버스 정류장, 대기장소, 휴게소 등등)을 가지는 다양한 형태의 작은 구조물(sheds)들은 중앙의 녹지를 둘러싸고 있다.
이 구조물들은 얇은 철판으로 만들어졌고 다양한 유형의 지붕을 가진다. 접힌 벽은 테이블과 벤치로 사용될 수 있다. 그래서, 이 구조물은 주변의 상점(카페, 편의점, 슈퍼마켓 등등)의 일부로 사용된다.
다양한 패턴은 아스팔트 위에 새겨진다. 이 패턴을 따라 줄서기, 모이기, 행위는 휴먼스케이프의 확장된 패턴을 생성시킨다.

HUMANSCAPE

Shinminatomura (a cooperative program of Yokohama This is a proposal for Kumamoto station west exit square competition.
East exit square (54,000 users a day) is surrounded by large scale commercial facilities putted on super blocks. By contrast, On the other hand, West exit square (18,000 users a day) adjoins small scales composed of small shops and existing residential zone. So, west exit square It is frequently used by local residents. Furthermore, existing railways being located at the east side of this station, Shinkansen line faces to the west side. This It means that people come from other prefectures see the view of the West exit square at first. We proposed a seamless landscape combining small scales and the station building with human activity.
Overlapping of canopies formed like plants, sheds, floor patterns and users continues to the townscape.
Various form sheds having different functions (bus stop, waiting room, rest house and so on) surround a central green area.
These sheds are made of thin steel plates, and putted various types of roofs on. Folded Folding wall can be used as tables and benches. table and bench.
So, these sheds structure are used as a part of surrounding shops (cafes, convenience stores, grocery supermarkets and so on). Various patterns are described inscribed on the square asphalt ground.
Queuing, gathering, acting along these patterns generates additional expanded patterns of humanscape.

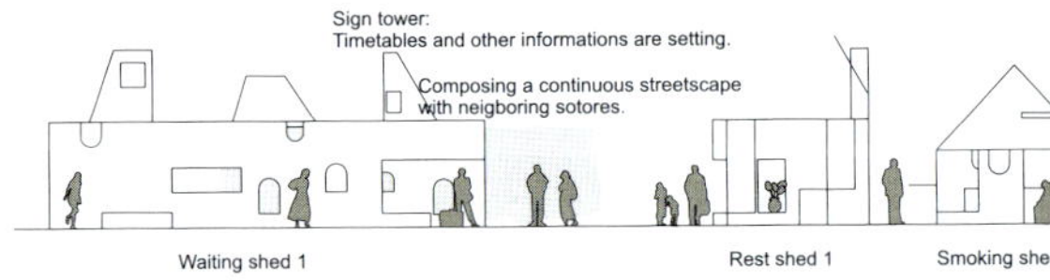

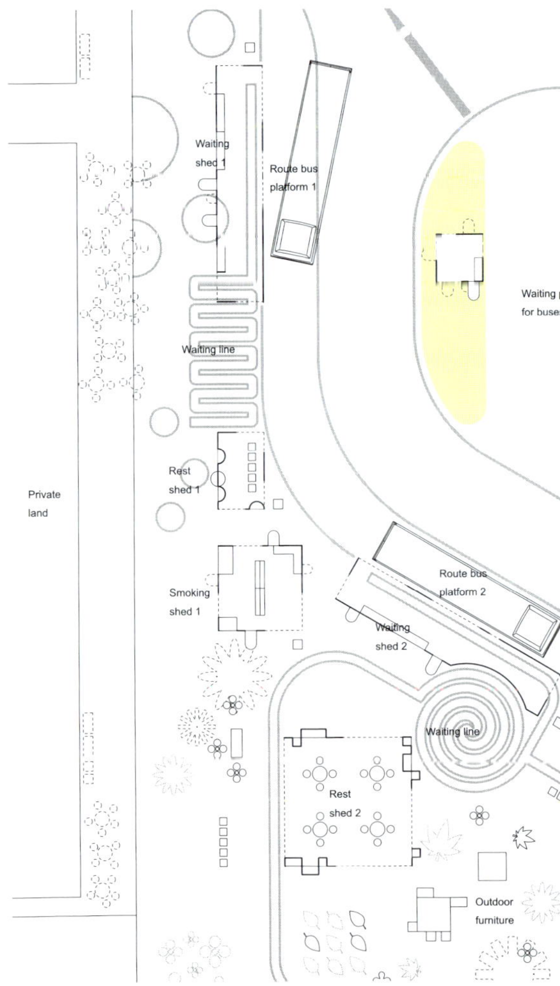

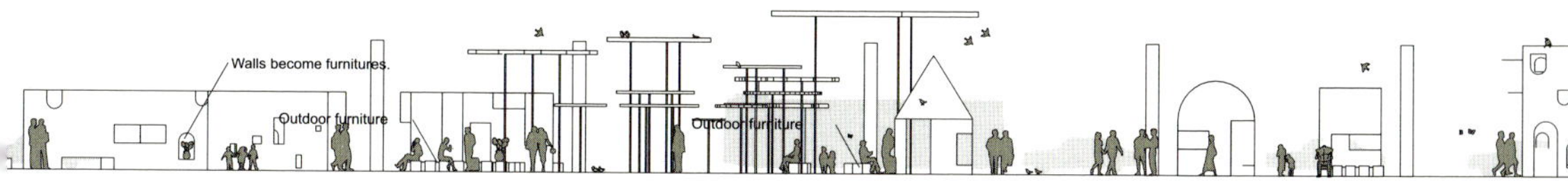

elevation 1/400

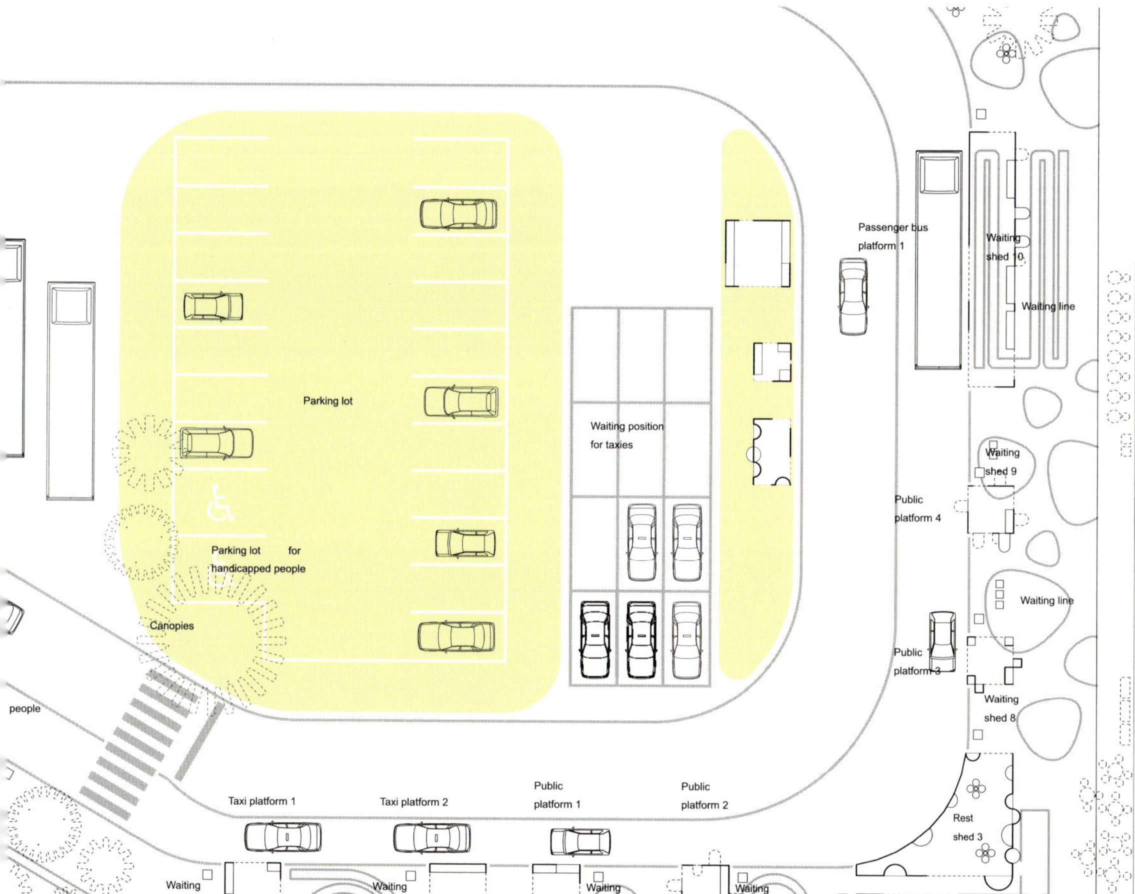

siteplan 1/400

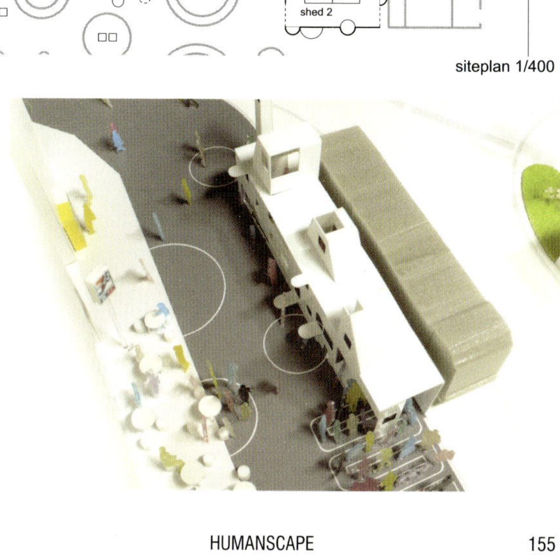

KU CO-OP
KU CO-OP
KU CO-OP
KU CO-OP
KU CO-OP

D-APARTMENT

1112

location:	Kadoma-city, Osaka
principal use:	apartment
site area:	161.72m^2
building area:	94.96m^2
total floor area:	228.33m^2
structure:	S
number of stories:	+3
completion:	December, 2011

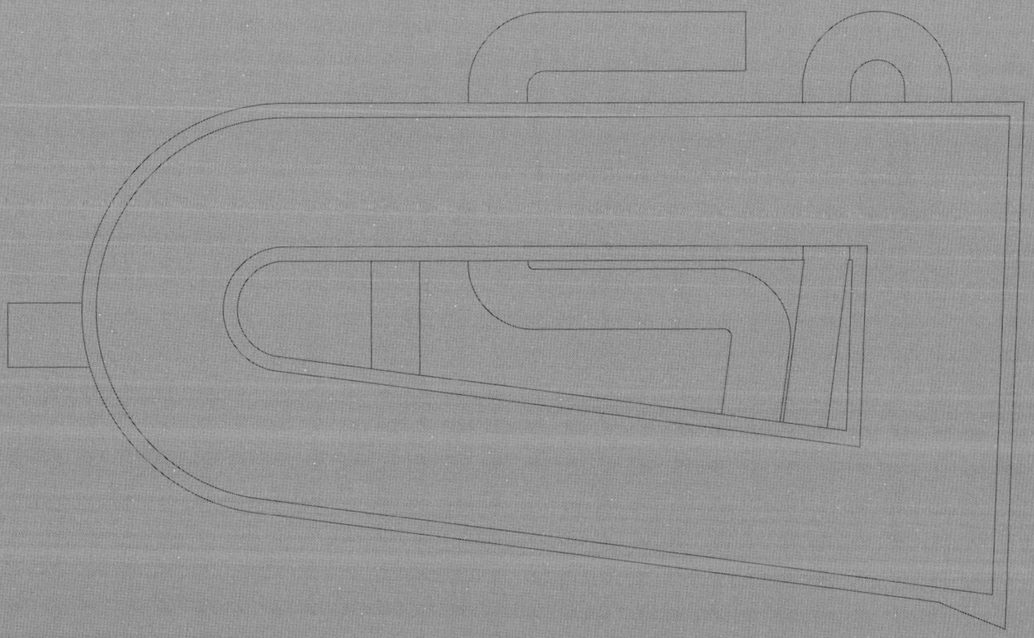

CONTENTS

유도하는 커뮤니케이션

공존의 방식"은 여러 사람들이 다른 사람들과 협력하여 함께 살아가는 건축적인 방법을 의미한다. (물론, 비-소통적인 방식도 때때로 필요하기는 하다.) 즉, 건축과 사회 사이의 관계를 규정하는 건축적 방법을 의미한다.

우리가 살고 있는 도시들은 복합적인 시스템이라고 할 수 있다. 물리적인 도시 환경은 구조물과 물류로 구성되어 있으며 비-물리적인 정보환경은 기후적 자연 환경으로 규정된다. 이러한 환경들은 서로 중첩되고 간섭된다. 우리 프로젝트들의 공통적인 모습은 이와 같은 세 가지 환경들에 접근하는 것이다. (자연환경, 도시환경, 정보환경)

각각의 프로젝트는 물리적인 도시환경과 자연환경(이것은 정보기술을 통해 보다 잘 시각화될 수 있다)에 대한 반작용(reaction)으로서 계획되었고, 동시에 프로젝트에 의해서 만들어진 새로운 환경 전체로부터의 일련의 연속적인 작용(action)으로 계획되었다.

나아가, 몇몇 프로젝트에서는 사회 그 자체를 지원하는 비-물리적인 정보환경을 통한 커뮤니케이션도 디자인되었다. 건축 그 자체뿐만 아니라 가구, 옥외 시설, 그리고 천과 같은 것들을 사용함으로써 우리는 다양한 종류의 커뮤니케이션을 유도하는 유연하고 열린 디자인을 제안한다.

"The way of being together" indicates the architectural method that many people can live together in collaboration with others.(Of course, non-communication method is often required.) That is, the architectural method for defining the relationship between architecture and society.

The cities we live in are complex systems that the physical city environment consisted of structures and logistics and the non-physical informational environment are described on the climatically natural environment. These environments overlap and interfere with each other. The common appearance of our projects is accessing these 3 environments. (natural environment, city environment, informational environment)

Each project is made from thinking it as a reaction to the natural environment (It is more visualized by information technology) and the physical city environment, and thinking simultaneously a series of actions from the whole new environment made by the project.

Furthermore, the communication with the non-physical informational environment supporting the society itself is designed in some projects. By using furniture, outdoor facility and fabric not only architecture itself, we propose the gentle and open design that induces many kinds of communication.

and SS, the key problem of architectural representation is, to reflect or reveal the makings of the spatial container itself, rather than to experiment other kinds of symbolic or artistic representation. Emphasizing both inward and outward communication of buildings, OZ and SS are still trying to create architectural spaces where more distinctive communicative co-existence can take place. The differences in their attitude and strategies, however, cannot be overlooked and, in evaluating their communicative modes, the question of appropriateness or decorum still remain: in what level or to what degree will the formal experiments be executed? And, will which outcome be more likely welcomed and sustainable by its users?

shapes, colors, and materials, and thus makes us focus instead on volumes, lighting, and views, ultimately creating in economical and efficient way dynamic spatial composition in limited sources. Of the postmodern language, the elitist one (sophisticated use of wit and irony) and popular one (simple signs and sensual expressions), SS seems to lean towards to the latter approach in a mitigated way which is as easy as the message the designs convey.

In comparison, OZ's approach is more rationalist and propagandistic. Their expressed aim is to replace "closed-up, filled-out spaces of discontinuation" with an "opened-up spaces for communication" in which people could enjoy greater availability. To OZ, the lack of communal life in current housings is because of the economically driven planning and the lack of visions for better communal lifestyles as well. So, OZ pays much of their intellectual efforts to break the existing economic and legal constraints, to gain more compactly furnished space. OZ questions not only about how to get as much usable space as possible, but how to compress various functions into a limited space as densely and effectively as they can. OZ tries to create such spaces through overlapping, juxtaposing, or interlocking of dwelling units, often resulting in such acrobatic forms as public platforms in the Namyangju Townhouses and the layered public spaces in the Munjeong Public Rental Housing Project.

The public space is critical in OZ's strategies because, according to OZ, it is a shared space where both intended and accidental meetings would happen. The public space designed by OZ often contains the private and individual functions such as storage, views outward, gardening places, etc., becoming "transitional" or in-between places where individual lives extend outwards. The idea that shared public spaces can be potentially appropriated into private spaces is more persuasive in cases like Munjeong Project where individual living units are built in the most minimal of sizes and standards. The way OZ treats privacy is more conventional than SS in that OZ's in-between spaces containing condensed private functions results from cautious protection of core individual privacy.

OZ's architectural representation emphasizes the existence of these public spaces, which are often more prominent, brighter, and transparent, ultimately contributing to represent an "extruded community and its communal life." To both OZ conditions and programs. First, each project of SS shows an understanding of a design as a reaction to environment comprised of overlapping layers of urban, informational, climate environments. The multiple responses to these layers operate in various aspects in the whole design process, not just on the surface of the building but all parts of the building, inside and out. By modifying conventional building types, SS keep trying to enhance, from their response to individual site conditions and programs, greatest possibilities such as utmost usable volumes, utmost lighting, utmost variety, etc., and is quite adept at inventing unconventional spatial forms, such as a long and winding room, an extremely tall room, a stepped garden, a parking-plaza, structure-furniture, etc. The collage-like collection of such metamorphic forms, displayed at "BankArt Life 3 Project at Yokohama Triennale," paradoxically proves how well they operate only in real on-site situation.

It is necessary to remember that for this invention SS had to reinterpret or negate to some degree the factors sustaining the existing spatial types; SS has proven their strength in flexible renovations by flexible design in the most conventional forms of privacy. One of their trademark designs is to expose private living area to its adjoined neighborhood or street, sliding on the borderlines of privacy: in "Ground and above roof house" (2010), its almost transparent living room flanks the street at grade, and in "Re-Formed Garden House" (2012), its garden is openly visible by its neighbors, and most dramatically, in "D-Apartment" (2011), its living units face one another around a narrow void. This loss of privacy is regarded by SS, inversely, as a maximized possibility of communication.

Any blind pursuit for newness might hinder the possibilities of communication or impede on comfort and convenience of dwelling, but SS managed to set new forms of living in a new architectural container, through so called "soft communication," which we may call the best part of their making communicative architecture. SS mitigates the impacts from sharp changes in the form of living, not only through smartly using informational technology, but also, more importantly, through properly selecting various elements, even from urban settings or furniture, so that everybody can easily understand their design, as in the case of Isoyama Dispensing Pharmacy (2010) where the tone-downed subtle graphic elements are applied to enhance communication. Furthermore, their preferred whitewashed mono-toned interior performs several related functions in communication: making the interior as clean space as possible, it unites the elements of various

RECOMMENDATION | 추천의 글

Architecture as Communicative Devices

SEO Jeongil (Seoul National University)

Presented here in comparison, the recent works by Design Group OZ and SPACESPACE ('SS'), mostly comprised of residential building projects closely related to our everyday life, tell us that even in an apparently globalized world of today, different cultural contexts still persist in the level of everyday life between Korea and Japan. So, in order to truly differentiate their works, one should know the different realities of both cultures. Yet, we can also find here a shared theme, "the ways of being together," which is admittedly faithful to both of their design approaches. Both firms believe that people should live together in communicative and cooperative ways, and that communicative architecture which furthers the collaboration and understanding between people can be gained by design, though they differ significantly in their actual interpretation and design.

Not new is the issue of communicative architecture: rather, it has been repetitively raised in the modern history of architecture, resulting various ways of understanding and solutions. So, to judge the works of OZ and SS, would be necessarily to contextualize their positions among these modern precedents. Yet their approach is in some aspects new and worth while to analyse. First, it is notable that both firms try to reflect, in their design, the local neighborhood dimension of everyday life. Their works maybe closely reflect current building businesses and policies, as, for example, the recently boosted business of small housing projects by governmental policies in Korea influenced OZ's practice. OZ and SS may be improving the quality of living conditions in some ways, although, not linked to any planning tool, their improvements remain within the already given conditions. The other reason why their approaches are interesting is that it is in the dwelling architecture that they are inducing communication, including man-to-man contact between residents, which is a tricky and delicate task since the place for communication today is primarily public places and private places, rather than private, residential buildings.

Certainly, OZ and SS's works are well above the average in many way in finding clear logics from the relation between site

우스), 공용 복도와 공용실의 레이어(문정동 공공원룸임대주택) 등에서처럼 그 결과물은 종종 곡예 같은 공간다이어그램으로 나타난다.

OZ의 디자인 전략에서 특히 공용공간이 핵심인데, 그 공용공간은 우연적, 의도적 만남이 일어나도록 의도된 공유공간이다. 이와 동시에, 이 공용공간은 개별공간에서는 여건상 모두 충족되지 않는 사적기능들-수납, 외부전망, 정원 등-이 거기에 적절히 배치될 때 그 공간에 사적 활동이 부분적으로 연장되어 "전이 공간" 혹은 사이공간으로 작용하게 된다는 것이 OZ의 논리다. 사적 전용의 여지를 둔 공유공간의 개념은 원룸임대주택처럼 개별 사적공간이 최소화된 주거유형의 경우 더욱 설득력 있어 보인다. 분명한 점은, OZ는 SS와 달리 자신의 주거공간의 가능성을 넓히는 시도에서 프라이버시를 희생시키는 일에 아주 조심스럽다는 점이다. OZ는 개별거실을 외부에 노출시키게 하거나 하지 않고, 공유공간에 빠져나온 사적요소들이란 개별적 삶에서도 비교적 프라이버시의 침해문제가 덜 한 요소들이다. 결국, 그 전이공간은 주거공간을 더 압축적으로 만들되, 관습적 형태의 프라이버시는 보수적으로 보호하려는 시도의 결과다.

OZ에게는 그 공용공간의 존재를 드러내는 것이 건축적 표현에 핵심이 되고 있다. 즉, 음각된 밝은 투명함으로 표현된 공용공간 부분은 "집합적 삶은 음각으로 개별적 삶은 양각으로 빚어진 압출성형된 공동체"를 표현하는 것을 돕게끔 의도되었다. OZ의 건축에서는 이미 프라이버시의 수준에 따라 공간들이 분절되어 있기 때문에 SS처럼 부드러운 소통의 문제가 덜하고 따라서 좀 더 강한 표현들을 드러내고 있다. 하지만, 대체로 OZ나 SS나 모두 건축적 표현이란 주거공간 또는 삶의 공간적 "용기"(容器)를 SS처럼 그대로 외부로 내어놓거나 강조해서 제시하는 것에 다름 아니다. 의식했건 아니건, 미적 문제, 또는 상징적, 예술적 재현의 차원에서 소통을 도모하고 있지 않고, 용기의 제작 또한 그것의 직접적 표현에 거의 전적으로 기대고 있다.

OZ와 SS의 소통장치들은 그 지향은 전반적으로 비슷해 보일지라도 각각의 태도와 전략은 각 지역의 문화적 차이만큼이나 편차가 있다. 소통 문제의 해법을 보여주기 가장 어려운 건축유형인 주거를 통해 둘 다 거기서 건축물의 바깥을 향한 소통과 안을 향한 소통 모두를 모색하고 있다. 그런데, 그들의 건축에서 특정한 소통의 국면이 더 잘 의도되어 있고 잘 드러나 보인다 해도, 다른 조건들과의 얼마나 균형 잡는가의 문제, 즉, 적절성(데코룸)의 문제가 남는다. 즉, 얼마나 적절한 수준에서 그 형태실험을 해 나아가느냐의 문제다. 결국, OZ와 SS의 디자인의 차이는 적절성의 감각을 어떤 부분에 얼마나 발휘하며 소통의 강도를 조절하느냐의 차이다. 역사적 선례가 보여주듯이 소통의 가능성은 OZ와 SS 사이의 진폭, 그리고 그 너머에서 걸쳐 있을 것이고, OZ의 보다 추상적이고 합의적인 소통도 SS의 보다 개성적 소통도 둘 다 우리 공동체의 공존방식의 하나일 수는 있다. 하지만, 당분간 어떤 방식이 사람들에게 더 잘 수용되고 어떤 방식이 더 지속될지?

정보기술을 포함한 기술을 재치 있게 사용하는 것—예를 들어, 지상 · 옥상의 집(2010)의 경우처럼 도 SS가 새로운 변화가능성을 넓히는데 있어서 중요한 수단이다.

전복을 위한 전복, 가능성을 위한 가능성은 소통을 이끌어내기보다 그 반대로 거주성을 훼손할 수도 있고 거주자의 불편한 적응을 불러일으키게 된다. 이를 피하고 새로운 삶을 건축물 안에 보다 안정정으로 안착시키는 능력이 SS에게 있다면, 그것은 그들이 건축유형의 변화에 따른 충격을 이른바 '부드럽고 열린 소통' 의 방식으로 완화시키는 방식에서 두드러진다. 다시금, 정보기술도 그런 역할을 하지만, 무엇보다, 건물의 구조적 요소, 가구, 옥외의 각종 기물과 가구 등의 다양한 요소들로부터 누구나 쉽게 이해할 수 있는 요소들을 선별해 적절히 구사하는 방식이 그 부드러운 소통과 관계된 듯한데, 가령, 이소야마 조제약국(2010)에서 보듯이 건축에 적용된 누그러뜨려진 미세한 수퍼그래픽이 그 건물에서 핵심적인 소통수단으로 삼고 있다.

여기에 더해, SS의 백색 단색조 내부공간은 최대한 밝고 위생적인 주거공간을 만드는 동시에, 재료, 색, 식재에 의한 다채로운 표현을 억제하여 형태와 공간, 시각패턴, 빛과 시선에 대한 관심으로 집중하게 만든다. 이것은 다양한 공간구성을 경제적으로 구사하는 백색 모더니즘의 전형적 특징이다. 하지만, 이와 달리, SS의 건축에는 대중적 포스트모더니즘의 접근방식들이 배어있다. 즉, 포스트모던 건축은 위트와 아이러니를 세련되게 구사함으로써 다소 엘리트주의적인 경향도 있었지만, 대개는 대중들이 이해하기 쉽도록 간명한 기호와 감각적 표현을 구사했고, SS는 이 후자의 접근을 취한다. 더구나, SS가 인간의 삶에 대해 전하려는 내용도 그만큼 쉽다. 아마 이런 접근들로 인한 '부드러운' 소통의 효과들을 믿기 때문에 SS는 "주택에서 도시를 누릴 수 있게 해 줄 것이며, 나아가 건축이 지금까지 다루지 않은 다양한 것들을 다루게 해 주고, 건축을 주변의 도시와 더욱 잘 연결해 줄 것"이라고 믿는 것일 테다.

OZ 역시 환경과 프로그램을 의식하고 반응하는 건축을 추구하지만, OZ의 접근방식은 비언어적이고 감각적 측면보다 논리적 측면이 우세하다. 즉, OZ는 '막히고 채워진 단절의 공간'을 '트인 공간'으로 대체해서 거기서 사람들이 더 여유 있고 서로 더 잘 만나도록 하는 디자인을 고안하는 것을 뚜렷이 목적으로 삼고 있다. OZ는 이 진단을 특히 공동주택에 적용해 왔는데, 그들은 현재의 공동주택에서의 공동체생활이 매우 미약한 수준에 놓여있는 이유에 대해 부분적으로는 경제적, 법적 제약 때문으로 부분적으로는 더불어 사는 삶에 대한 상상력이 부족함 때문으로 보는 것 같다. 보통의 경제적 조건과 법규로 인해 제약된 공간적 한계를 개별 사적공간이건 공용공간이건 우선 여유공간을 한 뼘이라도 더 확보하는 데 지적 노력을 동원하여 가용면적을 확보하면서 그 공간이 복수의 기능을 가지는 '압축적' 공간이길 OZ는 원한다. 이를 위해 OZ는 개별주호의 중첩, 병치, 맞물림 등을 실험하면서 교묘한 공간배치를 고안하는데, 예를 들어, 옥상과 지하의 공공플랫폼(남양주 타운하

주거환경의 기본적 질을 일정 정도 개선하는데 기여하고 있다.

다른 한편, 일상적 주거유형에서 소통을 진작하는 문제란 사실 어려운 문제다. 특히 구성원끼리의 대면소통을 문제 삼을 때 그렇다. 더구나, 우리가 일반적으로 경험하듯이, 오늘날 공적 장소 또는 소통의 장소는 사적 장소 또는 프라이버시의 장소로부터 대개 분리되어 있다. 소통은 일차적으로 공공건축과 공공장소의 문제지 프라이버시가 중시되는 섬세한 건축유형인 주택에서는 다루기 어려운 것이다. 하지만, SS도 OZ도 대지의 제약조건과 프로그램 간의 관계로부터 가능한 간명한 설계 논리를 도출하면서 소통을 시도하고 있는 중이다. 그것은 얼마나 유효한 물음이고 어디까지 추구되어야 바람직한 것일까?

어쨌든 OZ와 SS는 주어진 환경과 프로그램으로부터 도출할 수 있는 최대의 가능성을 논리적으로 도출하는 능력에서 여러 동세대 건축가들을 앞서가고 있다. 우선 SS의 경우, 그들은 한 건물이 처한 환경을 길과 건물의 도시환경, 비물질적 정보환경, 기후적 자연환경, 이 셋의 다층적으로 중첩된 환경으로 보고, 이에 반응하는 건물을 목적으로 삼는다. 그 반응의 모색 과정은 디자인 전과정에서 실현되고 있고 그 반응의 결과는 건물 외피 같이 건물의 특정 장소에서만 작동하는 게 아니라 건축물 안팎에 골고루 퍼져 있다. SS는 각 프로젝트마다 가진 고유한 대지조건과 프로그램의 특징에 주목하여 기존건축유형을 최대한 변형하는 데 관심을 두는데, 이 변형의 일차적 목적은, 최대한의 가용공간, 최대한의 채광조건, 최대한 변화 있는 내부공간 등을 확보하는 데 있다. 그 결과 예외적인 기본형태들 즉, 굽은 방, 긴 방, 높은 방, 계단형 정원, 주차장 겸 소광장, 구조체 겸 가구 따위를 얻어내게 되는데, SS는 자신들의 작업에서 실현된 이 이례적 형태들을 요코하마 트리엔날레에 설치한 뱅크아트 라이프 3 서점(2011)에 콜라주로 전시한 적 있다. 이렇게 모아 놓은 형태들은 아이러니하게도 그것들이 현장에 놓여 있는 경우와 비교하면 설득력이 훨씬 덜 하며, 따라서 거꾸로 SS의 이형적 형태들이 얼마나 현장의 맥락성에서 잘 작동하는지를 증명한다.

기억할 점은, 이 이형적 형태들을 만들어 내기 위해 기존건축유형을 지속시키는 여러 요인들을 재해석하거나 부정해야 했다는 점이고, 이점은 특히 관습적 형태의 프라이버시에 대해서 두드러진다. SS는 '유연한' 사고방식으로 '유연한' 구성을 이끌어 내기 위해 사적 거주공간을 개방시킬 수 있는 거의 한계까지 외부로 노출시키곤 하는데, 예를 들어, '지상과 지붕위의 집'(2010)에서는 거실을 1층 길에 바로 면하게 해서 거실이 마치 숍이나 카페 같이 길에 노출시켰다. 또한, 나라 이코마시의 주택(庭の形, 2012)에서는 마당을 이웃에 대해 과감하게 노출시켰다. 무엇보다 가장 극적인 사례는 D-아파트(2011)로서, 여기서는 폭이 3미터도 안 되는 좁고 긴, 꺾이고 굽은 개별 주거단위공간들의 양쪽에 창을 냄으로써 그만큼 폭이 좁은 안쪽 '중정' 보이드를 통해 각 거주자가 서로 넘겨다보이도록 허용했다. 이런 식으로 거주자의 프라이버시에 대한 관습적 입장을 깬 것에 대해 SS는 프라이버시의 양보라기보다는 확대된 소통의 기회로 삼을 수 있음을 주장한다.

소통장치로서 건축

서정일(서울대인문학연구원 HK연구교수)

설계집단 오즈와 스페이스스페이스(이하 각각 OZ, SS)의 최근작들–가장 기본적 일상환경을 이루는 건축유형들–을 나란히 놓고 보면, 한국과 일본이 세계화과정을 통해 많이 동질화되어있을 성싶어도 일상수준에서는 그 문화적 맥락의 차이가 여전히 뚜렷이 존재함을 새삼 깨닫게 된다. 그러니 그 문화적 맥락을 자세히 알고서야 두 사무소에 의한 작업의 차이도 제대로 이해할 수 있을 것이다. 하지만 여기서 OZ와 SS는 공존방식이라는 공통주제를 제시하며 자신들의 건축을 현대건축이라는 하나의 동시적 과정으로 봐 주길 제안하고 있는데, 이 공통주제는 그들 각자의 작업방식을 분명 어느 정도 충실히 반영하고 있고 그들은 분명 어느 수준에서 공유점을 지니고 있는 것 같다. 즉, 개별적으로 닫혀 제각기 존재하는 것보다 열린 방식으로 상호반응하고 공유하고 협력하는 공동체와 그것을 위한 건축을 원한다는 수준에서는 목표가 크게 다르지 않다. 이 목표 아래 그들이 실현하려는 건축은 곧 소통을 진작하는 건축이기에 그들에게 건축은 기본적으로 '소통장치'인 셈이다. 그 소통이 건축물을 매개로 해서 사람들 간에 직접적으로 불러일으키던, 건축물의 표현을 통해 간접적으로 불러일으키던 말이다.

하지만, OZ와 SS가 도시공동체의 상호소통에 기반을 둔 공존을 내세우고 있다하더라도, 그 구체적 해석과 건축적 결과물은 꽤 다르다. 역사적으로 볼 때, 그들이 제기하는 소통 문제 자체는 새로운 것은 아니나. 즉, 건축과 도시에서 공존, 소통, 결국 의미확보의 문제는 건축에서의 모더니즘의 역사에서 자주 제기되었고, 특히 20세기 후반이후의 중심 문제 중 하나로서 다양한 이해방식과 해결책을 낳았다. 그래서 결국 두 사무소의 작업을 궁극적으로 판명하는 것은 다양하게 존재하는 우리 삶의 공존과 소통에 대한 입장들 중에서 또한 현대건축의 역사적 선례와 비교했을 때, 그 위치가 어디에 있는지 판단하는 일이다.

OZ와 SS가 소통의 주제를 도시일상환경을 이루는 소규모 단독주택, 공동주택, 근린생활시설의 수준에서 진지하게 다루는 사실은 몇 가지 점에서 흥미롭다. 우선, 이 사실이 상당히 반가운 측면은, 그것이 최근의 주택시장 및 정책의 변화를 반영하기 때문인데, 한국만 해도 부동산경제와 주택정책의 효과로 지난 몇 년간 다세대 · 다가구 · 도시형생활주택의 건설이 크게 늘었으면서도 계획상의 획기적 개선은 없었던 상황을 OZ의 작업은 고스란히 반영한다.
이들 유형의 건물들은 그 동안 건축디자인의 관심 밖으로 밀려나 여전히 계획과 디자인상에서 새로운 돌파구를 못 찾고, 인구급증에 따른 높은 주거밀도 비좁은 주거면적, 불충분한 채광 · 환기 · 소음환경, 부족한 공용공간과 공공공간의 문제 등 대도시주거환경의 문제를 미루어 왔다. 이 상황에서 OZ와 SS의 세대가 그것을 진지한 디자인의 고려대상으로 다룬 시도는 그만큼 반가운 것이다. 비록 그 두 사무소 모두가 주거밀도와 환경조건을 바꾸는 적극적인 계획수단은 구사하지 못한채 주어진 계획과 정책의 조건 안에서 가능한 개선의 여지들만 디자인의 영역에서 발견하는 한계 안에서 작업했지만, 그들은

THE WAYS OF BEING TOGETHER

INDUCING COMMUNICATION

스페이스스페이스
SPACESPACE

共存
方式

SPACE

SPACE